Digital Innovations in Architecture, Engineering and Construction

The Architecture, Engineering and Construction (AEC) industry is experiencing an unprecedented transformation from conventional labor-intensive activities to automation using innovative digital technologies and processes. This new paradigm also requires systemic changes focused on social, economic and sustainability aspects. Within the scope of Industry 4.0, digital technologies are a key factor in interconnecting information between the physical built environment and the digital virtual ecosystem. The most advanced virtual ecosystems allow to simulate the built to enable a real-time data-driven decision-making. This Book Series promotes and expedites the dissemination of recent research, advances, and applications in the field of digital innovations in the AEC industry. Topics of interest include but are not limited to:

- Industrialization: digital fabrication, modularization, cobotics, lean.
- Material innovations: bio-inspired, nano and recycled materials.
- Reality capture: computer vision, photogrammetry, laser scanning, drones.
- Extended reality: augmented, virtual and mixed reality.
- Sustainability and circular building economy.
- Interoperability: building/city information modeling.
- Interactive and adaptive architecture.
- Computational design: data-driven, generative and performance-based design.
- Simulation and analysis: digital twins, virtual cities.
- Data analytics: artificial intelligence, machine/deep learning.
- Health and safety: mobile and wearable devices, QR codes, RFID.
- Big data: GIS, IoT, sensors, cloud computing.
- Smart transactions, cybersecurity, gamification, blockchain.
- Quality and project management, business models, legal prospective.
- Risk and disaster management.

Nazly Atta

Green Approaches in Building Design and Management Practices

Windows of Opportunity Towards Circularity

Springer

Nazly Atta
Department of Architecture, Built
Environment and Construction Engineering
(DABC)
Politecnico di Milano
Milan, Italy

ISSN 2731-7269 ISSN 2731-7277 (electronic)
Digital Innovations in Architecture, Engineering and Construction
ISBN 978-3-031-46759-2 ISBN 978-3-031-46760-8 (eBook)
https://doi.org/10.1007/978-3-031-46760-8

This Springer imprint is published by the registered company Springer Nature Switzerland AG
The registered company address is: Gewerbestrasse 11, 6330 Cham, Switzerland

Paper in this product is recyclable.

Preface

The book deals with the themes of Green Transition and Circular Economy in the fields of Building Design, Built Environment and Real Estate management.

Circularity represents a hot topic today in many business fields, promoted and supported by various European Union (EU) regulations, directives and initiatives. In particular, the building and Real Estate sector is considered by EU a key priority area also with respect to the achievement of the United Nations (UNs) Sustainable Development Goals (SDGs), especially the Goal 12 on sustainable production and consumption.

Replying to the emerging requirements set by EU and UNs, the diffusion of the concepts of Circular Economy and circularity within the building sector has gradually consolidated the awareness that extending the life of products represents a winning strategy for pursuing the optimization of the use of materials and the reduction of environmental impacts. Re-strategies (reuse, remanufacturing, reconditioning, repurpose, etc.) are promoting multiple uses of goods, bringing the attention to the end of service-life of building elements and reviewing the concepts of building maintenance and service provision.

The key stakeholders of the building and Real Estate sector currently recognize the increasing importance of integrating circular practices within design and management processes in order to guarantee a consequent wise management of the resources embedded in buildings during their whole useful life, reducing waste generation and performing an efficient use of resources.

Although the expected benefits arising from these circular approaches, the common practice within the building sector is still characterized by linear processes, also due to the absence of shared sector references and easy-to-apply guidelines on circularity issues able to support stakeholders (including Real Estate owners, designers, construction companies, building managers, service developers, maintenance operators, facility managers, waste managers, etc.) in integrating circular practices within their businesses.

Indeed, building and Real Estate operators are expressing a growing demand of new methodological tools and procedural frameworks to support them in understanding the nature and the features of these circular approaches, as well as of new strategies and standards in order to become promoters of circularity within the sector, seizing the opportunity to comply with European environmental sustainability policies, regulations and initiatives.

Therefore, by outlining some windows for seizing the opportunity to integrate circular approaches in current practices, the book aims to provide information tools to support building and Real Estate stakeholders in:

- understanding re-strategies and the related requirements and preconditions for their implementation within building design and management practices;
- reviewing current design approaches at the product, process and building scale;
- developing new circular organizational and contract models for the procurement of building products and Facility Management (FM) services;
- drawing new supply chains based on network approaches;
- assessing the performance, effectiveness and efficiency of (implemented/to implement) circular practices.

In particular, the book is organized into five chapters, followed by Conclusions.

Chapter 1 introduces the theoretical framework and State of Art of the topic of green and circular transition within the building and Real Estate sector, highlighting key concepts, paradigm shifts and trends.

Chapter 2 introduces three key "windows of opportunity" for integrating circular approaches within building practices along the Building Process, as well as the related enabling role of stakeholder networks in the achievement of circularity integration.

Chapter 3 focuses on the design stage, proposing new project approaches towards Design-for-Deconstruction/Reuse/Remanufacturing at the product, building and service level, highlighting new circularity-related contents for the Briefing Documents (BDs) according to the indications introduced by the EU Level(s).

Chapter 4 focuses on the management stage, proposing new circular formulas for the supply of building products and Facility Management (FM) services according to EU guidelines on Circular Procurement. Accordingly, new contents for properly expressing circularity-related requests within Invitations to Tender (ITTs) for building products and FM services provision are introduced.

Chapter 5 focuses on the assessment stage, understanding the impending obligations for building operators and Real Estate companies deriving from the recent EU regulatory updates on Sustainability Reporting, including new disclosure requirements on circularity performance. Hence, emerging needs and potential "game changers" for a compliant and efficient reporting of circular building practices are highlighted.

Lastly, the author proposes a glossary, useful for the better understanding of the terminology adopted in this book and, at the same time, to summarize the meaning that relevant terms belonging to the Circular Economy field acquire when specifically applied to the building and Real Estate sector.

Milan, Italy Nazly Atta

Acknowledgements

The work presented in this book is part of the Research "Remanufacturing: strategies for extending the life of building products. New design approaches, innovative manufacturing and organizational models and circular processes" (2022–2024, ongoing), conducted at the Department of Architecture, Built Environment and Construction Engineering (DABC) of Politecnico di Milano, supported by eFM SpA and co-funded by the European Union within the National Operational Programme (NOP) Research and Innovation 2014–2020—additional ESF REACT-EU resources, Action IV.6 Research contracts on green topics, Ministerial Decree no. 1062 of 10 August 2021 of the Italian Ministry of University and Research (MUR).

Contents

Chapter 1
Circular Re-strategies in Building Design and Management: Reviewing Basic Concepts

1.1 Green Versus Disposal: Introducing Resource-Resource Models

In recent years, the concepts of Circular Economy (CE) and Green Transition have been widely deepened and experimented in several business sectors, aiming at optimizing the use of materials and reducing environmental impacts. The different CE approaches and definitions (Table 1.1) proposed by literature and by the several experimentations within different business sectors—with particular reference to industrial fields—have already demonstrated the interest for circular practices from both academics and professionals.

As confirmed by literature contributions and professional practices (Eberhardt et al. 2022; Cruz Rios et al. 2021; Terzioğlu 2021; Rios and Grau 2020), the interest is focused on circular approaches based on the so-called "Re-strategies" (Table 1.2)—i.e. reuse, reconditioning, remanufacturing and repurposing—as defined in the BS 8887-2 standard "Design for manufacture, assembly, disassembly and end-of-life processing (MADE). Part 2: Terms and definitions" by the British Standards Institution (BSI).

Among the various industries, the European Commission recognizes the key role of the construction sector, which currently represents a "priority area" due to its significant impact on the environment. According to the Eurostat statistical data, in 2018 the construction industry contributed for 35.9% of the total waste generated by all economic activities and households in Europe (Eurostat 2021). Morevoer, Ness and Xing (2017) stated that globally the construction industry accounts for 33% of greenhouse gas emissions, 40% of resource consumption, and 40% of waste generation (Ness and Xing 2017; Rahla et al. 2021).

In this regard, the operators of the building and Real Estate sector consider Circular Economy and the paradigm of circularity as suitable responses to this resource-intensive field (Hossain et al. 2020; Sáez-de-Guinoa et al. 2022), fostering sustainability in a systematic way and overcoming traditional linear models based

N. Atta, *Green Approaches in Building Design and Management Practices*,
Digital Innovations in Architecture, Engineering and Construction,
https://doi.org/10.1007/978-3-031-46760-8_1

Table 1.1 Definitions of the Circular Economy (CE) concept

Reference	Circular Economy definition	Keywords
Preston (2012, p. 1)	"*Circular economy is an approach that would transform the function of resources in the economy. Waste from factories would become a valuable input to another process – and products could be repaired, reused or upgraded instead of thrown away*"	Repair; reuse; upgrading
Ellen MacArthur Foundation (2013, p. 7)	"*A circular economy is an industrial system that is restorative or regenerative by intention and design. It replaces the 'end-of-life' concept with restoration, shifts towards the use of renewable energy, eliminates the use of toxic chemicals, which impair reuse, and aims for the elimination of waste through the superior design of materials, products, systems, and, within this, business models*"	Restorative and regenerative systems; product design for waste reduction
EEA (2014, p. 11)	Circular economy "*refers mainly to physical and material resource aspects of the economy – it focuses on recycling, limiting and re-using the physical inputs to the economy, and using waste as a resource leading to reduced primary resource consumption*"	Recycling; re-using; waste as a resource
COM (2015) 614, European Commission (2015, p. 2)	Circular economy is an economy "*where the value of products, materials and resources is maintained in the economy for as long as possible, and the generation of waste minimised*"	Waste minimization; extension of product value overtime
Sauvé et al. (2016, p. 49)	Circular economy refers to the "*production and consumption of goods through closed loop material flows that internalize environmental externalities linked to virgin resource extraction and the generation of waste (including pollution)*"	Closed loop material flows
Van Buren et al. (2016, p. 3)	"*A circular economy aims for the creation of economic value (the economic value of materials or products increases), the creation of social value (minimization of social value destruction throughout the entire system, such as the prevention of unhealthy working conditions in the extraction of raw materials and reuse) as well as value creation in terms of the environment (resilience of natural resources)*"	Economic, social and environmental value of products; resilience
Murray et al. (2017, p. 377)	"*The Circular economy is an economic model wherein planning, resourcing, procurement, production and reprocessing are designed and managed, as both process and output, to maximize ecosystem functioning and human well-being*"	Conscious design of the resourcing, procurement and reprocessing activities

Table 1.2 Definition of key Re-strategies by the BS 8887-2 standard (BSI)

Re-strategy	Definition by BS 8887-2:2009 (BSI)
Reuse	According to the BS 8887-2 standard, the term "reuse" refers to a set of operations "*by which a product or its components are put back into use for the same purpose at end-of-life*" (BS 8887-2:2009).
Reconditioning	The term "reconditioning" is defined by the BS 8887-2 standard as the "*return a used product to a satisfactory working condition by rebuilding or repairing major components that are close to failure*" (BS 8887-2:2009). With respect to the reconditioned element, its overall performance is likely to be inferior to that of the original element, as the extension of its warranty.
Remanufacturing	As defined by the BS 8887-2 standard, "remanufacturing" refers to the "*Return a used product to at least its original performance with a warranty that is equivalent or better than that of the newly manufactured product. Note 1: from a customer viewpoint, the remanufactured product can be considered to be the same as the new product. Note 2: manufacturing effort involves dismantling the product, the restoration and replacement of components and testing of the individual parts and whole product to ensure that it is within its individual parts and whole product to ensure that it is within its original design specifications*" (BS 8887-2:2009). The performance of the remanufactured element is expected to be at least as the one of the original element (original performance specification) and its warranty is generally at least equal to that of new product.
Repurposing	The term "repurposing" refers to the use of "*a product or its components in a role that it was not originally designed to perform*" (BS 8887-2:2009). Hence, the element undergoes transformative processes, resulting into another element with a different function from the original one.

on the "extract-use-landfill" approach (Hossain and Thomas 2019; Rahla et al. 2021) preferring circular organizational models based on resource reuse and recovery.

Hence, the issues of circular economy and ecological transition have animated debates on a national and European scale in the last decade, gradually consolidating the awareness that extending the life of products represents a winning strategy for pursuing within the building and Real Estate sector the key sustainability objective of optimizing the use of materials and products while achieving a reduction of the environmental impacts. In particular, the basic Circular Economy concepts of waste prevention and resource reuse and recovery (Fig. 1.1) have been already successfully implemented in various fields, ranging from machinery to electrical equipment and mechanic industry, however their application within the building sector is still experimental, having a shorter history and a slighter extent (Norouzi et al. 2021), essentially limited to recycling and material processing (Adams et al. 2017; Ghisellini et al. 2018; Di Biccari et al. 2019) rather than adopting a holistic approach (Rahla et al. 2019; Minunno et al. 2020).

In this regard, several recent policies and regulations of the European Union (EU) demonstrate the strong interest in the implementation of circular strategies based on reuse and remanufacturing practices in the building and Real Estate sector,

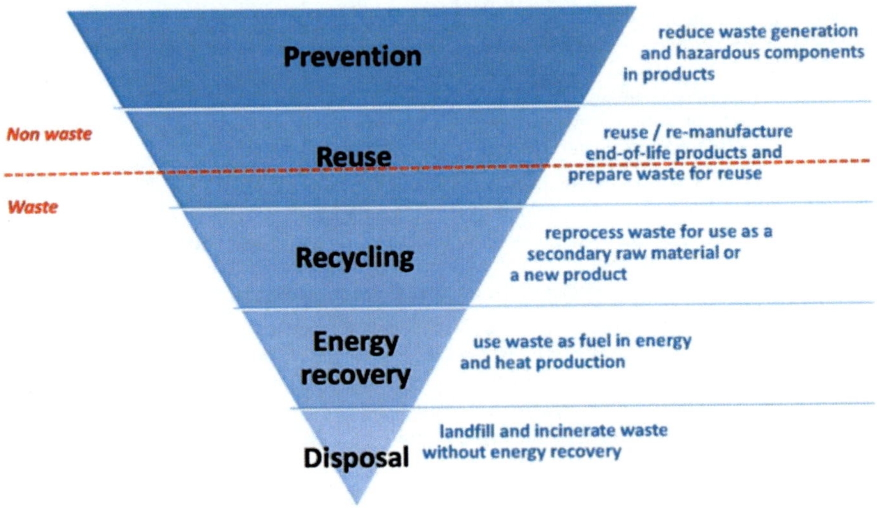

Fig. 1.1 Circular Economy strategies: from waste disposal to waste prevention. *Source* OECD (2020)

promoting circular initiatives and offering support tools to operators towards the pursuit of circular objectives. In particular:

- the Directive 2008/98/EC[1] establishes the commonly recognized "waste hierarchy", promoting support measures for the definition of sustainable production and consumption patterns, also encouraging the design, production and use of durable, "re-workable" and reusable products. In particular, the Directive establishes the guidelines for waste treatment at the European level, defining a "waste hierarchy", meant as a series of strategies—ordered by priority—to be implemented in order to manage waste with the lowest possible environmental impact, namely: (a) prevention; (b) preparing for reuse; (c) recycling; (d) other recovery, e.g. energy

[1] The Directive 2008/98/EC of the European Parliament and of the Council on waste and repealing certain Directives "*lays down measures to protect the environment and human health by preventing or reducing the generation of waste, the adverse impacts of the generation and management of waste and by reducing overall impacts of resource use and improving the efficiency of such use, which are crucial for the transition to a circular economy*" (Directive 2008/98/EC). In particular, the Directive establishes a common legal framework for treating waste in the EU. The framework aims to protect the environment and human health by stressing the importance of proper waste management, recovery and recycling techniques to reduce and improve the use of resources. Moreover, the Directive sets some minimum operating requirements for extended producer-responsibility schemes, including the responsibility to contribute to waste prevention and to the reusability and recyclability of products. For what concerns the waste generation, the Directive calls EU countries to take measures to: support sustainable production and consumption models; promote the design, manufacturing and use of products that are resource efficient, durable, reparable, reusable and capable of being upgraded; encourage the availability of instruction manuals, technical information, spare parts or other means allowing the repair and reuse of products without compromising their quality and safety (Directive 2008/98/EC).

recovery; and (e) disposal. According to this order of priority, the primary objective is "prevention", i.e. the reduction of waste production. In order to reach this goal, the Directive promotes circularity-oriented design approaches, conceiving products in a way that favors the propensity to future disassembly, repair, reuse and remanufacturing activities, avoiding potential waste production during their life cycle.

- The Green Deal[2] proposes a "Circular Economy Action Plan" aimed at promoting a policy of "sustainable products" giving priority to reuse, in order to exploit the residual performance of products before recycling them, emphasizing the importance of new circular business models and new measures to encourage companies to offer, and consumers to buy, durable, reworked and/or reused products. In particular, the Plan document states the intention to establish a legislative initiative on "sustainable products" based on a set of "sustainability principles" that includes (European Commission 2020):

  - improving durability, maintainability, reusability and repairability of building elements, increasing energy efficiency and optimizing the use of resources;
  - increasing the recycled content of products, while guaranteeing their performance and safety;
  - limiting the single-use and the premature obsolescence;
  - encouraging the "Product-as-a-Service" approach and business models in which the manufacturer retains the ownership of the product and/or the responsibility for the product performance during its life cycle (extended producer responsibility);
  - implementing digitalization in the processes of gathering and managing product information including advanced solutions such as IDtags, RFID, IoT (Internet of Things), sensor networks and Digital Product Passports.

[2] The European Green Deal is a package of policy initiatives aimed at supporting EU in the path to the green transition towards climate neutrality. The Green Deal underlines the need for a holistic and cross-sectoral approach. Hence, it involves initiatives covering interlinked areas, including: climate, environment, energy, transport, industry, agriculture and sustainable finance (European Council 2022).

- Mandatory and voluntary regulatory standards at project and product scale, such as the Ecodesign Directive,[3] the Ecolabel[4] and the Green Public Procurement (GPP),[5] developed in past years, promote a transition towards "green" practices and the supply of sustainable, durable and reusable building elements.
- The most recent Level(s) framework[6] proposes a common European language for measuring efficiency and circularity in construction (Table 1.3) with the aim of enriching the system of metrics for assessing circularity of building projects, creating a common approach based on the integration of current certification tools (European Commission 2021a).

---

[3] The EU Ecodesign Directive—"*Directive 2009/125/EC of the European Parliament and of the Council of 21 October 2009 establishing a framework for the setting of Ecodesign requirements for energy-related products*"—establishes a framework for the integration of environmental aspects into the design of energy-related products. The aim of this preventive approach is to optimize the environmental performance of products while maintaining their functional qualities. Indeed, the Ecodesign Directive encourages manufacturers of energy-using products to reduce the energy consumption and other negative environmental impacts during the whole product life-cycle. To this end, the Directive sets some reference performance criteria to which specific product categories must conform to be put on the market. These criteria and measures includes specifications of requirements for eco-design, contents of technical documentation, reference parameters and benchmarks, procedures for conformity assessment and testing procedures for market surveillance purposes (Directive 2009/125/EC).

[4] The EU Ecolabel—"*Regulation (EC) No 66/2010 of the European Parliament and of the Council of 25 November 2009 on the EU Ecolabel*"—is a voluntary environmental labeling scheme promoted by the European Union to distinguish products and services which, while guaranteeing high performance, are characterized by a limited environmental impact during their entire life-cycle. EU Ecolabel is based on a system of selective criteria which takes into account the environmental impacts of products or services throughout their entire life cycle (Regulation EC 66/2010). EU Ecolabel allows consumers to make conscious choices without sacrificing product quality. By choosing the Ecolabel, on one side, consumers select products that have a high ecological quality, are certified by independent bodies (competent national bodies) and are recognized at European level. On the other side, operators (e.g. manufacturers, service providers, resellers, etc.) gain visibility on the market for their commitment to the environment, contributing to the shift towards a "green" market (ISPRA 2022).

[5] Green Public Procurement (GPP) is defined in the Communication COM (2008) 400 "Public procurement for a better environment" as "*a process whereby public authorities seek to procure goods, services and works with a reduced environmental impact throughout their life cycle when compared to goods, services and works with the same primary function that would otherwise be procured.*" (COM 2008:400). Specifically, the term "Circular procurement" refers to an approach to GPP which pays special attention to "*the purchase of works, goods or services that seek to contribute to the closed energy and material loops within supply chains, whilst minimizing, and in the best case avoiding, negative environmental impacts and waste creation across the whole life-cycle*" (European Commission 2017).

[6] The Level(s) framework is based on six macro-objectives that address key sustainability issues throughout the building life-cycle. The framework is based on six macro-objectives, tracked through sixteen indicators. The latter describe how the building performance is aligned with the strategic EU policy objectives in areas such as energy, water, indoor air quality material use and waste, and resilience to climate change (European Commission 2022).

Table 1.3 Level(s) circularity-related indicators—macro-objectives 2 (resource use) and 6 (costs and value). Adapted from: European Commission (2021b)

	Objective		Description	Indicator categories	
Level/s framework	2	Resource efficient and circular material life cycles	Optimize building design to support lean and circular flows, including: – Minimization of construction and demolition waste generated to optimize material use – Replacement cycles and flexibility to adapt to change – Potential for deconstruction (opposed to demolition)	2.1	Bill of quantities, materials and life-spans
				2.2	Construction and demolition waste and materials
				2.3	Design for adaptability and renovation
				2.4	Design for deconstruction, reuse and recycling
	6	Optimized life cycle cost and value	Long term view of whole life costs and market value of buildings, including: – Life cycle costs (construction, operation, maintenance, refurbishment, disposal) – Integration of sustainability aspects into market value assessments and risk rating processes	6.1	Life cycle costs (€/m^2/yr)
				6.2	Value creation and risk factors

- The EU Taxonomy[7] for sustainable activities, which represents a key element of the EU sustainable finance framework. The EU Taxonomy aims to push financial and industrial sectors towards investments for climate neutrality by defining the conditions that construction activities have to meet in order to be considered "sustainable" and, thus, be eligible for sustainable investments. To this end, the EU Taxonomy establishes a set of classification criteria for defining "sustainable economic activties", also focusing on the construction sector—encompassing new building construction, building renovation, maintenance interventions, etc. (Table 1.4). Moreover, the accompanying Do-No-Significant Harm (DNSH) standards[8] involve circular economy practices including reuse, recycling or material recovery of construction waste, pollution minimization, and biodiversity protection through environmental impact assessments (PEEB and GlobalABC 2022).
- On a more general level, strategic programmes such as "Next Generation EU"[9] and "National Recovery and Resilience Plans"[10] consolidate the growing interest of European public institutions in promoting circular economy, also underlining the strong link between two key themes of the green transition: sustainability and digitalization.
- The 2030 Agenda for Sustainable Development, adopted by all United Nations (UN) Member States in 2015, proposes 17 Sustainable Development Goals (SDGs) (Fig. 1.2), representing an urgent call for action by all countries (UN 2015). In particular, the construction sector is commonly considered as a key area to meet the UN SDGs (Nußholz and Milios 2017; Sáez-de-Guinoa et al. 2022), especially with reference to the Goal 12 on ensuring sustainable production and

[7] The Action Plan On Financing Sustainable Growth—within the action category "Reorienting capital flows towards a more sustainable economy"—called for the creation of the EU taxonomy, meant as a common classification system for environmentally sustainable economic activities. The EU taxonomy sets a list of economic activities and it establishes for each of them a set of conditions that they have to meet in order to be qualified as "environmentally sustainable". The Taxonomy Regulation establishes six environmental objectives: (1) Climate change mitigation; (2) Climate change adaptation; (3) Sustainable use and protection of water and marine resources; (4) Transition to a circular economy; (5) Pollution prevention and control; (6) Protection and restoration of biodiversity and ecosystems (Regulation EU 2020/852).

[8] The DNSH principle is based on the EU Taxomony for a sustainable finance, adopted to promote private sector investment in green and sustainable projects and support to achieve the goals of the Green Deal. The EU regulation on Taxonomy of sustainable activities—as per EU Regulation 2020/852—introduces a set of DNSH criteria for determining how organizations contributes significantly to environmental protection, without causing significant harm to any of the six environmental objectives (see footnote 7) (Regulation EU 2020/852).

[9] *Next Generation EU* is a EU economic recovery package to support the EU member states to recover from the COVID-19 pandemic. It supports innovation and sustainable development towards climate, green and digital transitions. The key instrument of the *Next Generation EU* is the *Recovery and Resilience Facility* (RRF), which provides grants and loans to support reforms and investments in the EU Member States (European Commission 2021c).

[10] A National Recovery and Resilience Plan represents a roadmap for key reforms and projects, outlining objectives and investments that a EU Member States intends to carry out through NextGenerationEU funds to mitigate the COVID-19 socio-economic impact.

Table 1.4 Examples of activities included in the EU taxonomy belonging to the construction sector—category of activities no. 7 "Construction and Real Estate activities". *Source* Regulation EU 2020/852, Annex 1

	Type of activity by EU Taxonomy		Description	Classification reference/s
7. Construction and real estate activities	7.1	Construction of new buildings	Development of building design and construction projects of residential and non-residential buildings for later sale	NACE codes F41.1 and F41.2, including also activities under F43, in accordance with the statistical classification of economic activities established by Regulation (EC) No 1893/2006
	7.2	Renovation of existing buildings	Construction and civil engineering works or preparation thereof	NACE codes F41 and F43 in accordance with the statistical classification of economic activities established by Regulation (EC) No 1893/2006. An economic activity in this category is a transitional activity as referred to in Article 10(2) of Regulation (EU) 2020/852
	7.3	Installation, maintenance and repair of energy efficiency equipment	Individual renovation measures consisting in installation, maintenance or repair of energy efficiency equipment	NACE codes F42, F43, M71, C16, C17, C22, C23, C25, C27, C28, S95.21, S95.22, C33.12 in accordance with the statistical classification of economic activities established by Regulation (EC) No 1893/2006

(continued)

Table 1.4 (continued)

	Type of activity by EU Taxonomy		Description	Classification reference/s
	7.5	Installation, maintenance and repair of devices for measuring and controlling building energy performance	Installation, maintenance and repair of instruments and devices for measuring, regulation and controlling energy performance of buildings	NACE codes F42, F43, M71, and C16, C17, C22, C23, C25, C27, C28, in accordance with the statistical classification of economic activities established by Regulation (EC) No 1893/2006
	7.6	Installation, maintenance and repair of renewable energy technologies	Installation, maintenance and repair of renewable energy technologies, on-site	NACE codes F42, F43, M71, C16, C17, C22, C23, C25, C27 or C28, in accordance with the statistical classification of economic activities established by Regulation (EC) No 1893/2006
	7.7	Acquisition and ownership of buildings	Buying real estate and exercising ownership of that real estate	NACE code L68 in accordance with the statistical classification of economic activities established by Regulation (EC) No 1893/2006

consumption patterns, which also includes the improvement of the efficient use of natural resources (Goal 12.2), the prevention and reduction of waste generation (goal 12.5) and sustainable public procurement (12.7).

All these tools, representing a support towards the green transition, have stimulated the operators of the building and Real Estate sector to experiment the development and application of new circular models based on life extension strategies of building products and components, embracing the "resource-resource" approach and going beyond recycling. The following paragraph introduces these virtuous circular models that allow to shift from the linear "take-make-dispose" model to a circular economy, maintaining the value embedded in building elements as long as possible, thus minimizing resource use and waste generation through the implementation of repair, reuse and remanufacturing practices.

Fig. 1.2 The 17 Sustainable Development Goals (SDGs) by United Nations (UN). *Source* UN website (https://www.un.org/sustainabledevelopment/blog/2015/12/sustainable-development-goals-kick-off-with-start-of-new-year/#)

1.2 Pay-for-Use Versus Pay-for-New: Reshaping Value Creation

In the context of the experimentation and application of circular economy practices, strongly promoted by the EU regulations and initiatives introduced in the previous paragraph, a decisive element is represented by the definition of circular organizational models, understood as structured sets of strategic, tactical and operational activities through which the creation of value takes place (Table 1.5).

In this perspective, analyzing recent literature contributions (Mackenbach et al. 2020; Charef and Emmitt 2021; Ogunmakinde et al. 2022; Azcarate-Aguerre et al. 2022), referring both to research and professional activities, some innovative approaches emerge within the building sector aimed at extending the useful life of products through innovative "service-oriented" approaches (Rios and Grau 2020). These approaches are based on the new concept of "servitization", which implies a paradigm shift towards practices no longer oriented towards the sale of the product but towards offering the product "as-a-service". According to this new vision, the client goes from being a "consumer" of a product to becoming a "user" who pays for the service through "pay-per-service unit" formulas, including pay-per-use, pay-per-period or pay-per-performance formulas (Table 1.6). In this way the ownership of the product is not transferred to the client but it remains in the hands of the provider, who has extended responsibilities with respect to the performance of the product. Hence, the shift from product-centric to service-centric organizational models is realized.

Table 1.5 Linear versus circular model: key differences

	Linear economy model	Circular economy model
Step plan	Take–make–dispose	Cradle to cradle
Cycle	Open	Closed
EoL strategy/ies	Disposal	Reduce-reuse-recycle
Resource use	Mass use and consumption of new raw materials	Exploitation of the residual value embedded in already-existing products, minimizing the use of new raw materials
Waste	Waste generation	Waste prevention
System scope	Short term. From purchase to disposal	Long term. From purchase to reuse
No. of use cycles	1 (single use)	Multiple subsequent use-cycles
Organizational model	Focused on products	Focused on services
Value creation	Mass production and product selling	Selling as-a-service new and reused/ remanufactured/repurposed products (collecting them after-use)

Table 1.6 Key features of "Pay-per" formulas

Pay-per formula	Features
Pay-per-service unit	"Pay-per-service unit" is a remuneration model according to which the payment depends on the quantity and/or quality of service that the client receives. "Pay-per-service unit" includes several sub-categories such as pay-per-use, pay-per-period and pay-per-performance.
Pay-per-use	The "Pay-per-use" formula implies that the client pays on the base of what it effectively consumes. Hence, in this model, the actual usage of a service or product is metered.
Pay-per-period	The "Pay-per-period" formula refers to the payment of a variable fee by the client according to the extension of the timeframe of use of the service or product. Hence, in this model, the duration of availability of a product and/or service to the client is measured.
Pay-per-performance	The "Pay-per-performance" formula, also called "performance-related payment", refers to the variability of the economic amount of the payment with respect to the actual performance delivered to the client. In this model, the performance results are monitored through a set of Key Performance Indicators (KPIs). By tying the service price to performance, "pay-per-performance" formulas encourage the providers to supply high quality products and/or services to their clients.

The latter are no-longer based on a one-time delivery of a product but on a continuous performance delivery, focusing on reuse to increase the number of use-cycles and on maintenance to extend the lifetime of products.

In this regard, on a European scale it is possible to observe a progressive introduction of new practices and new organizational and business models for the building sector based on: (i) maximizing the cycles of use of products and (ii) implementing preventive maintenance strategies to extend products useful life, guaranteeing their functional and economic performance. In particular, the following circular organizational models emerges from literature and case study reviews (Charef and Emmitt 2021; Ogunmakinde et al. 2022; Azcarate-Aguerre et al. 2022):

- "Product-as-a-service" or product-service models. According to these models the client is not the owner of the asset but it pays for the access to its use "as-a-service". The remuneration is measured and delivered through innovative pay-per-use or pay-per-period formulas (Table 1.6). In this way, the client only pays for how much/long it uses the physical good;
- renting and leasing models. These models involve the payment of a periodic fixed "fee" for the temporary use (according to variable periods of time) of the good, which is returned at the end of the pre-established period in order to start a subsequent new cycle of use. The client pays the fee at pre-established regular time-intervals (e.g. monthly, every six months, etc.) independently from the intensity of the use;
- "shared-use" models. These models are based on the sharing (even simultaneously) of a physical asset by various users, maximizing the use of the good itself and allowing users to reduce the access costs to the asset;
- "all-in" models. These models are based on the so-called "all-inclusive formula", hence the purchase by the client of the physical asset together with a set of support services (including assistance and ordinary maintenance) useful for increasing the reliability of the asset itself, guaranteeing its performance over time and enabling multiple use cycles. This kind of models often involves a deposit-based formula according to which the user pays for the physical good plus an extra share of deposit to be returned once the use of the asset ends. When the good is returned (termination of the property), the deposit is returned to the user.

These innovative circular models based on the use of the physical asset are closely linked to the concepts of reliability, availability, maintainability and serviceability (Table 1.7). In fact, throughout the entire time-period of use of the element, its performance and functionality must be guaranteed with high levels of quality.

A support in this direction is represented by the most recent solutions of Information and Communication Technologies (ICTs) and, in particular, "Internet of Things" (IoT) and Big Data management. Such digital data management technologies, and the related advanced ICT-based data-driven services (Table 1.8), are able to significantly support the development of organizational and business models based on the innovative concept of "servitization".

The combination of circular practices with the use of digital tools for advanced information management represents a keystone in the process of transition towards circularity, contributing to maintain—even over several use-cycles—the added value of the material and energy resources integrated into the building and its parts, reducing the use of new resources and the impacts on the environment. In this regard, the

Table 1.7 The "abilities" of circular products and services

Ability	Definition
Reliability	The concept of "Reliability", as defined by the ISO/IEC 25010 standard refers to the "*degree to which a system, product or component performs specified functions under specified conditions for a specified period of time*" (ISO/IEC 25010:2011). Moreover, the standard ISO/IEC 2398-14 defines Reliability as "*the ability of a functional unit to perform a required function under given conditions for a given time interval*" (ISO/IEC 2382-14:1997)
Availability	The concept of "Availability" refers to the "*degree to which a system, product or component is operational and accessible when required for use*" (ISO/IEC 2510:2011)
Maintainability	The term "Maintainability" is defined by the ISO/IEC 2382-14 standard as "*the ability of a functional unit, under given conditions of use, to be retained in, or restored to, a state in which it can perform a required function when maintenance is performed under given conditions and using stated procedures and resources*" (ISO/IEC 2382-14:1997)
Serviceability	The term "Serviceability" is defined by the ISO/IEC 2382-14 standard as "*the ability of a service to be obtained at the request of the user, and to continue to be provided for a requested duration, within specific tolerances and under given conditions*" (ISO/IEC 2382-14:1997)

following paragraph introduces the main innovative methods and tools—with particular reference to ICTs—towards a circular knowledge management within traditional building practices.

1.3 Circular Versus Linear Information Management: Boosting Process Innovation

A shift from linear to circular Information Management (IM)[11] processes is required in order to implement the circular organizational models described in the previous paragraph (see Sect. 1.2). Indeed, if the processes of managing data and information are linear it is not possible to achieve circularity in the management of building elements. As highlighted in Table 1.9, if—on one hand—linear information management involves a one-way linear progression of data management, circular information

[11] Information Management (IM) is here meant as the management of organizational processes, activities and systems that acquire, create, organize, store, distribute, and use data and information. In particular, according to Jaeger et al. (2005), Information Management (IM) can be defined as "*the entire range of technical, operational, and social functions of a system that is used to handle information. [...] Information management affects the organization of information, the access to information, and the ways in which users can interact with the information. [...] Information management is usually influenced by a strategy or framework that guides the planning, application, and uses of that information. [...] Information management strategies influence the relationships between information, technology, and the larger social or organizational information, and have an impact on the functions of information and information systems*" (Jaeger et al. 2005).

Table 1.8 ICT-based services to support circular organizational models based on "servitization"

ICT-based service	Added value within circular models	Involved tools	Benefits
Real-time assistance	This ICT-based service solution involves a set of Help Desk and Technical Support Services, including real-time technical assistance. These advanced services may be part of renting or leasing models or shared-used models, already included in the related "pay-per-use/period" payment formulas. They can be also part of "all-in" models as after-sale support services.	– Remote monitoring networks (based on sensors and smart devices) – Smart products (with embedded sensors) – Real-time communication system, including real-time online support (email, chat, etc.)	Among the benefits of the realtime assistance it is possible to mention: – Improved provider-client/users communication; – Prompt management of users requests for assistance/intervention; – Increased capabilities of fault detection and resolution; – Reporting and storage of feedback information useful for performing analyses on product failure profiles (failure type and frequency).

(continued)

Table 1.8 (continued)

ICT-based service	Added value within circular models	Involved tools	Benefits
Condition-based maintenance	Condition-based maintenance exploits the use of information technology to identify incipient failures before they occur, allowing a cost-effective and accurate planning and scheduling of preventive maintenance interventions. For the implementation of this strategy, it is necessary to define threshold values or acceptability ranges for a set of key parameters capable of describing the behavior/aspect of interest of the element (e.g. equipment, technical and plant engineering element, furnishing component, etc.). When these thresholds are exceeded, the system warns the provider to activate maintenance interventions in order to prevent the failures.	– Remote monitoring networks (based on sensors and smart devices) – Data visualization tools – Sensing and Responding systems – Platform for data analytics	Condition-based maintenance allows to: – Guarantee reliability and availability by determining the need for maintenance activities based on the real conditions of the building elements; – Optimize the maintenance program by scheduling only the necessary interventions and when required, in order to optimize maintenance costs while making the program effective (scheduling of the needed interventions at the right time, preventing breakdowns); – Prepare the necessary knowledge base to conduct predictive analyses on future degradations/failures of the elements. Using predictive data analysis techniques, based on the monitoring of conditions, usage status and degradation/failure profiles of the building elements, it is possible to establish the most probable "time to failure" of each element, hence improving the capability to plan maintenance and act proactively.

(continued)

Table 1.8 (continued)

ICT-based service	Added value within circular models	Involved tools	Benefits
Re-strategy digital simulation	The development of a virtual representation of existing elements (digital twins) mirroring the physical and technical characteristics and behavioral performance of the real physical elements overtime allows to digitally simulate possible re-strategies (e.g. reconditioning, remanufacturing, repurposing, etc.) assessing their technical and economic feasibility (e.g. propensity of the element to be easily dis/assembled; possibility to rework/substitute/modify/resize/ etc.).	– Remote monitoring networks (based on sensors and smart devices) – Digital twin – Data visualization tools	Reduced uncertainties on: – Product dis/assemblability assessments; – Choice of most suitable re-strategy; – Re-strategy cost estimation; – Re-strategy performance estimation; – Needed tools and skills to perform re-actions; – Future performance of the reworked product.

(continued)

Table 1.8 (continued)

ICT-based service	Added value within circular models	Involved tools	Benefits
Smart billing for use-based payment	Smart billing refers to the "metered billing" formula that includes the innovative "pay-per" pricing systems (e.g. pay-per-use, pay-per-period, pay-per-performance). According to these payment systems, the clients/users only pay for what they have used in each billing cycle. In particular, by means of smart sensors, meters and detectors (metering system), the company providing the product/service can track the usage of the product/service by users. The latter are only charged for what they have used—detected by the metering system.	– Remote monitoring networks (based on sensors and smart devices) – Data visualization tools – Service/Product Usage monitoring systems (metering systems) – Digital billing apps – Smart web/mobile interfaces	Smart billing enable to achieve a: – Better control on element/service availability and/or performance; – Greater flexibility in service offers; – Increased client/user awareness on product use and consumptions; – Transparent payment policy; – Information base on element/service use for each client useful to tailor the service/product offer.
Digital simulation for circular design	On the basis of the outcomes of the "re-strategy digital simulation" and taking into account the actual use of elements, it is possible to exploit virtual modelling software to digitally design re-usable/re-manufacturable/re-purposable elements. Indeed, using a digital representation it is possible to design elements according to specific requirements (modularity, dis-assemblability, etc.) that allow an easy and quick disassembly of the elements in their future useful-lives, facilitating reuse and multiple use-cycles. Moreover, it is also possible to check the effectiveness of the design-for-disassembly/design-for-remanufacturing strategies by running simulations of the breakdown/re-strategy processes on the designed elements.	– Digital 3D modelling software – Software to simulate the disassembly procedure (design-for-disassembly control)	Digital circular design and simulation of dis/assembly processes allow to: – Design circular elements with a high propensity to be reused; – Control the effectiveness of the "design-for-disassembly" and "desing-for-remanufacturing" strategies; – Implement circular models based on remanufacturing and reuse of building elements and furniture

management implies a continuous process of data collection and updating, including data reliability and validity assessments for the subsequent storing and processing (Table 1.9). The two approaches also differ in terms of data value creation strategy, indeed a linear IM process is characterized by a one-to-one relationship between data and value, namely: one data is used in one process to reach a single specific purpose. In such a linear process, data are collected on "as-needed" basis —e.g. only when needed to solve a problem or to check a performance. Instead, a circular process is characterized by one-to-many relationship between data and value, indeed key-data are periodically collected and shared among the different stakeholders within a unique common database (data sharing). Hence, according to a circular approach, data and information are integrated and they are of multiple-utility since they serve different actors and different processes, even at different times (data are stored in the common database and they can be extracted by different stakeholders with different purposes when needed). Thus, if the focus of a linear approach is on data effectiveness, the focus of a circular approach is also on data-efficiency, reached by implementing common shared procedures and tools for data collection, storage, sharing, monitoring and updating. In this regard, it is important to stress how in the context of circular IM processes is of fundamental importance to implement common templates, procedures and tools to collect feedback information (after-use/-intervention data) from different stakeholders in order to enrich the existing knowledge base, achieving a continuous improvement of processes.

Table 1.9 Linear versus circular Information Management: key features and differences

	Linear Information Management	Circular Information Management
Process features	Easy to scope and closed-ended	Open-ended and iterative
Step plan	One-way linear progression	Continuous process of collection, updating, storage, sharing and processing of data and information
Database	Vertical independent databases managed by different stakeholders according to a "silos" approach	Unique database, shared among all the involved stakeholder
Data value	One-to-one, single value. One data useful for one process. Data collected on an "as-needed" basis	One-to-many. One data collected one time can be useful for different processes with different purposes at different times
Focus	Data-effectiveness. Data to activate one-time linear process	Data-effectiveness and data-efficiency. Monitoring, updating and enriching the knowledge base over time
Feedback information	Not collected	Collection of feedback data and information for continuous improvement of processes and constant enrichment of the existing knowledge base

Hence, as a cycle of iterative processes, a circular IM is able to support the key learning activities, namely: identification of information needs; collection of data and information; organization and storage of data and information into databases; updating of data and information; analysis and interpretation of information; accessibility and dissemination of information; use and exploitation of data and information.

Hence, in the context of management of data and information within the building field, it is possible to define different connotations of circularity according to a threefold perspective, namely:

- circularity of procedures and tools, meant as the use—by the different stakeholders—of common information tools and procedures to collect, share and exchange data;
- circularity of information, meant as circularity of the data management process including data updating and collection of feedback information for a continuous enrichment of the knowledge base;
- circularity within stakeholder networks, meant as circulation of data, information, lesson learned and best practices among the involved actors.

Accordingly, circular information management processes in the field of building management are characterized by the following key attributes:

- ability to collect, integrate and link data and to interpret the resulting information base (data connection and interpretation);
- ability to access, disseminate and share information among the involved stakeholders (data accessibility and sharing);
- ability to extract meaning from data, hence to analyze current phenomena and to predict future ones (data structuring and analytics).

These needed abilities—attributes of a circular IM—have to be pursued encompassing at the same time: (i) the management of information resources, records, archives; (ii) the management of mandatory and voluntary policies, standards and references; (iii) the management of information processes, procedures and activities; (iv) the management of Information and Communication Technology (ICT) tools (Choo 2002).

In this regard, it is possible to highlight some preconditions (Table 1.10) for implementing the above-introduced abilities in order to achieve circularity of information management processes in the context of circular organizational models based on reuse and remanufacturing of building elements.

In particular, focusing on the role of information tools and the contribution of ICTs, it is possible to highlight the following supports:

- Implementation of a modular expandable information system or Information Platform (IP) and standardized procedures for the management of the multiplicity of heterogeneous data sources, as well as for the collection, normalization, integration, storage, sharing and analysis of data. This means that the IP covers at

Table 1.10 Examples of preconditions for circular IM processes in the context of building element reuse- and remanufacturing-based models

	IM interrelated aspects	Preconditions for a circular IM
Circularity—Circular IM	Management of information resources, records, archives	Procedures and tools for integration of heterogeneous data sources
		Procedures for data normalization and integration
		Procedures for classification and coding of building elements
		Procedures for classification and coding of documents
		Shared templates for collecting and organizing data and information concerning the building elements, both static and dynamic data and information. In particular: (i) static data, e.g.: technical data from manufacturers, dimensions, material, color, weight, procedures for dis-assembly, acquisition cost, etc.; (ii) dynamic data, e.g.: residual performance, maintenance needs, no. cycle of use, etc.
		Tools and procedures for data collection and storage overtime to outline the element history in terms of profile of use, performed use-cycles, received maintenance and/or remanufacturing interventions, occurred faults and/or anomalies, on-going and past use-contract, current and past owner/s, etc.
		Creation and updating of Product Passports and related attribution of authority and responsibility to Passport Developers and subsequent Passport Managers
	Management of mandatory and voluntary policies, standards and references	Definition of the mandatory and voluntary standard framework according to different scales (portfolio, building, element), different building stages (design, de/construction, maintenance and management) and different levels (strategic, tactical, operational). This process also includes regular checks for regulatory updates
		Definition of circularity-related mandatory requirements and consequent Information Need for the different phases of the building process (design, construction, maintenance and management, including reuse/remanufacturing/deconstruction)

(continued)

Table 1.10 (continued)

	IM interrelated aspects	Preconditions for a circular IM
		Definition of the required documents (both mandatory and voluntary) for the different phases of the building process at the building and element (product) scales, including for instance: product certifications, CE mark, eco-certification of environmental labels or environmental declarations, building certifications product warranties, maintenance manuals and plans, contractual clauses (e.g. for rent and leasing), etc.
		Definition of the validity duration of the documents and related deadlines and definition of the timeframe, information and procedures necessary for requesting the document updates (for documents that require the active involvement of other parties, e.g. authorized bodies and operators, public bodies, etc.)
		Allocation of the responsibility for the organization, assessment, storage and updating of documents (giving priority to the ones required to be in compliance with law) to one or more document managers
	Management of information processes, procedures and activities	Definition of use-based models (e.g. product-as-a-service, long-term renting, etc.) and related contractual terms and billing methods (e.g. pay-per-use, pay-per-period, etc.), information procedures (data extraction and updating from/to databases, data processing from monitoring systems, data aggregation for assessment dashboards, etc.) and related information tools
		Identification and characterization of the key stakeholders of the different processes (stakeholder analysis)
		Setting the rules of the reuse/remanufacturing supply network, including definition of roles, responsibilities, authorities, interests, boundaries, information-exchange obligations, contractual-relationships, etc. of the involved stakeholders
		Definition of information and communication strategies and development and dissemination among stakeholders of information and communication plans

(continued)

Table 1.10 (continued)

	IM interrelated aspects	Preconditions for a circular IM
		Definition of common tools, templates and procedures for collecting feedback information from stakeholders
		Definition of the needed skills for managing the information tools and services (information platforms, sensor monitoring systems, ICT- and IoT-based services, smart billing, etc.) and development of training plans for internal and external staff
	Management of information and communication technology (ICT) tools	Creation of a unique digital database to collect and share data, information and document according to common classification and coding rules. Allocation to the different stakeholders of suitable authorities to access/modify the contents of the database according to their specific role and purpose
		Implementation of a modular and expandable Information Platform (IP) based on the unique central database. The Information Platform acts at the same time as: (i) collector of information and documents from different sources; (ii) digital system to activate processes (including intervention requests, remanufacturing activities, activation/closure of contracts, etc.); (iii) system to perform reuse/ remanufacturing/repurposing feasibility analyses; (iv) digital marketplace for the demand and offer of reuse/remanufactured building elements
		Implementation of a digital tag system (e.g. RIFD, QRcode, etc.) of the elements for the collection and displaying of data (via mobile devices). The tag system has to be connected to the information platform to enrich and update the central database, guaranteeing the traceability of the building elements overtime

(continued)

Table 1.10 (continued)

	IM interrelated aspects	Preconditions for a circular IM
		Implementation of a sensor-network for monitoring the use of the elements and/or the offered service by different users. The sensor-network has to be connected to the Information Platform (possibility to add a module of the IP dedicated to the sensor monitoring) to enrich the information base, allowing to perform data analyses useful for, e.g. smart use-based billing, defining the profile of use of the different elements (basis for residual performance estimations), etc.
		Implementation of data visualization tools allowing the display of data through smart digital interfaces, e.g. dynamic dashboards including also graphic data representations, charts, etc. and the possibility to create an external *app* (connected to the IP) for the offer of information-based services (e.g. smart billing, request of interventions, displaying of contractual terms, etc.) to the users of the elements and supply-network stakeholder
		Data analytics for performing descriptive and predictive analyses, including for example: residual performance assessment, estimations of residual useful-life of elements, forecasting future degradations, faults or anomalies and activate maintenance interventions before their occurrence, estimation of no. of residual use-cycle, etc.
		Implementation of a 3D modelling software for the developments of digital twins of the building elements (enablers for the performing of simulations of residual performance for reuse, as well as dis-assembly and remanufacturing feasibility and sustainability). It is needed to check the interoperability with the central Information Platform and the sensor-based monitoring system

(continued)

Table 1.10 (continued)

	IM interrelated aspects	Preconditions for a circular IM
		Implementation of a specific module of the Information Platform to document management and, in particular, to Product Passport management. This document follows the building element throughout its whole useful-life and it becomes the key information tool for enabling multiple use-cycle (reuse/remanufacturing/ repurposing under the different introduced service- or product-based organizational models, see Sect. 1.2)
		Implementation of Blockchain technology for (i) transparency of processes within the client-provider relationships (e.g. billing systems, renting contracts, app for product-as-a-service, etc.) and (ii) tracing in a reliable and trustable way the chain of custody of the building element overtime

least the following macro-functions: document management, records management, web content management, digital asset management, learning management and analytics.

- Implementation of Digital Product Passports. A Digital Product Passport "*is intended to provide consistent track-and-trace information on the origin, composition, repair and dismantling options of a product, as well as on its handling at the end of its service life*" (Adisorn et al. 2021). Hence, it contains information such as the list of components, materials and raw materials used in the production of the building element, durability, upgradability, possibility of reuse/ remanufacturing or recovery of materials, information on how to disassemble the element or repair the parts most at risk of degradation and failure, expected generation of waste materials, historical and current data on the ownership and location of the building element, behavioral and use data from sensing devices, etc. This set of information is accessible by building operators in order to facilitate the maintenance, reuse, remanufacturing and repurposing of building elements or specific components.
- Exploitation of "Internet of Things" (IoT) for product tracking (traceability of building elements) (Atta and Talamo 2022; Atta 2023) as a tool for monitoring (even in real time) the use level of the different building parts (including technical elements, technological systems, equipment, furniture, etc.). By means of IoT it is also possible to monitor the state of use of shared assets (e.g. elements under a rent contract or a product-as-a-service formula) in order to make their availability/occupancy status visible remotely (e.g. via smartphone or via web) and to be able to book their use for a predetermined period of time. These tools also allow the creation of a "history" of use (traceability), useful for monitoring

the state of the elements over time, estimating their residual performance and scheduling any necessary preventive maintenance interventions.

- Implementation of a "Chain of Custody" framework based on Blockchain technology. A "Chain of custody" is defined as the "*chronological documentation that records the sequence of control, transfer, analysis, and disposition of evidence in a digital evidence system*" (Kao et al. 2018). In the context of circular organizational models, a Chain of Custody means developing a chronological documentation regarding a product (including product origin, past lifecycles, supply journey, usage billing, etc.), all the actors that come into contact with the product overtime (both on the supply and demand side) and the history of all the related transactions (even if occurred under different contractual formulas, e.g. direct sale, renting, as-a-service, etc.). Hence, the development of a Chain of Custody allows to increase the efficiency and transparency of circular processes, enhancing the whole accountability within the supply network. Moreover, by relying on Blockhain technology and its advanced serialization and authentication solutions it is possible to support the recording of information even through complex supply chain networks, by storing data in decentralized registries (Kouhizadeh et al. 2019) that are accessible—for visualize and filter data—by all the stakeholders involved in the circular transactional processes. In particular, "Blockchain" can be defined as a "*digital, decentralized and distributed ledger in which trans-actions are logged and added in chronological order with the goal of creating permanent and tamper-proof records*" (Treiblmaier 2018). The collection and storage of reliable data still represents a challenging activity in the building field, often difficult to perform. In this regard, such a technology can support the tracing of information in circular organizational models, which imply the activation of multiple sequential contracts (even with different clients) based on the use-reuse of the same product. Moroever, Blockchain solutions can be fed by the data collected by heterogeneous advanced technological solutions, including RFID (Radio Frequency IDentification), IoT (Internet of Things) smart devices and GPS (global position sensors), able to collect—even in real-time—accurate data, addressing data reliability and supply-chain traceability issues.
- Implementation of 3D modelling and behavioral simulation software, based on Building Information Modeling (BIM) (Liu et al. 2022) and Internet of Things (IoT), for the development of Digital Twins (DTs) of the building elements (Atta and Talamo 2022; Atta 2023). A digital twin is a virtual model that mirrors a physical entity in terms of physical, technical, behavioral and performance characteristics. The application of DTs to circular organizational models in the building practice represents an opportunity to: (i) increase the accuracy of the estimations of the residual performance (for reuse) of building elements by collecting overtime information on their usage-profiles; (ii) increase the estimation of the effort required to perform re-strategies (e.g. remanufacturing and repurposing) on building products/systems by performing digital simulations; (iii) improve product design by digitally assessing the ease of dis/assembly; etc. Hence, DTs make it possible to share relevant information among building stakeholders

(Liu et al. 2022) useful to assess different re-strategies and circular scenarios, thus supporting decision-making.

The integration of circular approaches and the above-mentioned necessary Information Management (IM) and Information Technology (IT) tools implies a review of traditional methods and tools for building design and management, also including: (i) the updating of current profiles of skills and competence of building stakeholders (designers, manufacturers, building managers, facility managers, etc.) and (ii) the updating of criteria and information tools for decision-making in the key stages of the Building Process. These aspects will be deepened in the next Chapter.

References

Adams KT, Osmani M, Thorpe T, Thornback J (2017) Circular economy in construction: current awareness, challenges and enablers. Proc Inst Civ Eng Waste Resour Manag 170:15–24

Adisorn T, Tholen L, Götz T (2021) Towards a digital product passport fit for contributing to a circular economy. Energies 14(8):2289

Atta N (2023) Remanufacturing towards circularity in the construction sector: the role of digital technologies. In: Arbizzani E et al (eds) Technological imagination in the green and digital transition. CONF.ITECH 2022. The Urban Book Series. Springer, Cham, pp 493–503

Atta N, Talamo C (2022) Advanced digital information management tools for smart re-manufacturing. In: Talamo C (ed) Re-manufacturing networks for tertiary architectures. Innovative organizational models towards circularity. Franco Angeli, pp 195–209

Azcarate-Aguerre JF, Andaloro A, Klein T (2022) Facades-as-a-service: a business and supply-chain model for the implementation of a circular façade economy. In: Rethinking building skins. Woodhead Publishing, pp 541–558

Charef R, Emmitt S (2021) Uses of building information modelling for overcoming barriers to a circular economy. J Clean Prod 285:124854

Choo CW (2002) Information management for the intelligent organization: the art of scanning the environment. Information Today, Inc.

Cruz Rios F, Grau D, Bilec M (2021) Barriers and enablers to circular building design in the US: an empirical study. J Constr Eng Manag 147(10)

Di Biccari C, Abualdenien J, Borrmann A, Corallo A (2019) A BIM-based framework to visually evaluate circularity and life cycle cost of buildings. IOP Conf Ser: Earth Environ Sci https://doi.org/10.1088/1755-1315/290/1/012043. Accessed December 2022

Eberhardt LCM, Birkved M, Birgisdottir H (2022) Building design and construction strategies for a circular economy. Arch Eng Des Manag 18(2):93–113

EEA (European Environment Agency) (2014) Resource-efficient green economy and EU policies. Publications Office of the European Union, Luxembourg. https://www.eea.europa.eu/publications/resourceefficient-green-economy-and-eu. Accessed December 2022

Ellen MacArthur Foundation (2013) Towards the circular economy. J Ind Ecol 2:23–44

European Commission (2017) Public procurement for a circular economy. Good practice and guidance. https://ec.europa.eu/environment/gpp/pdf/CP_European_Commission_Brochure_webversion_small.pdf. Accessed November 2022

European Commission (2020) Circular economy action plan. https://environment.ec.europa.eu/strategy/circular-economy-action-plan_en. Accessed November 2022

European Commission (2021a) Level(s): a guide to Europe's new reporting framework for sustainable buildings. https://ec.europa.eu/environment/eussd/pdf/Level_publication_EN.pdf. Accessed November 2022

European Commission (2021b) Level(s): putting whole life carbon principles into practice. Level(s) in practice series No. 2. https://op.europa.eu/en/publication-detail/-/publication/20761f2e-143f-11ec-b4fe-01aa75ed71a1/language-en/format-PDF/source-230076864. Accessed November 2022

European Commission (2021c) The EU's 2021–2027 long-term budget and NextGenerationEU. Facts and figures. https://op.europa.eu/en/publication-detail/-/publication/d3e77637-a963-11eb-9585-01aa75ed71a1/language-en. Accessed November 2022

European Commission (2022) How does level(s) work? https://environment.ec.europa.eu/topics/circular-economy/levels/lets-meet-levels/how-does-levels-work_en. Accessed December 2022

European Council (Council of the European Union) (2022) European green deal. https://www.consilium.europa.eu/en/policies/green-deal/. Accessed on December 2022

Eurostat (2021) Waste statistics. https://ec.europa.eu/eurostat/statistics-explained/index.php?title=Waste_statisticsTotal_waste_generation. Accessed November 2022

Ghisellini P, Ji X, Liu G, Ulgiati S (2018) Evaluating the transition towards cleaner production in the construction and demolition sector of China: a review. J Clean Prod 195:418–434

Hossain MU, Thomas NS (2019) Influence of waste materials on buildings' life cycle environmental impacts: adopting resource recovery principle. Resour Conserv Recycl 142:10–23

Hossain U, Ng ST, Antwi-Afari P, Amor B (2020) Circular economy and the construction industry: existing trends, challenges and prospective framework for sustainable construction. Renew Sustain Energy Rev 130:109948

ISPRA Istituto Superiore per la Protezione e la Ricerca Ambientale (2022) Ecolabel UE. https://www.isprambiente.gov.it/it/attivita/certificazioni/ecolabel-ue. Accessed November 2022

Jaeger PT, McClure CR, Thompson KM (2005) Social measurement in information management. In: Encyclopedia of social measurement. Academic Press, pp 277–282

Kao DY, Chao YT, Tsai F, Huang CY (2018) Digital evidence analytics applied in cybercrime investigations. In: 2018 IEEE conference on application, information and network Security (AINS). IEEE, pp 111–116

Kouhizadeh M, Sarkis J, Zhu Q (2019) At the nexus of blockchain technology, the circular economy, and product deletion. Appl Sci 9(8):1712

Liu Z, Wu T, Wang F, Osmani M, Demian P (2022) Blockchain enhanced construction waste information management: a conceptual framework. Sustainability 14(19):12145

Mackenbach S, Zeller JC, Osebold R (2020) A roadmap towards circularity-modular construction as a tool for circular economy in the built environment. IOP Conf Ser: Earth Environ Sci (IOP Publishing) 588(5):052027

Minunno R, O'Grady T, Morrison GM, Gruner RL (2020) Exploring environmental benefits of reuse and recycle practices: a circular economy case study of a modular building. Resour Conserv Recycl 160:104855

Murray A, Skene K, Haynes K (2017) The circular economy: an interdisciplinary exploration of the concept and application in a global context. J Bus Ethics 140(3):369–380

Ness DA, Xing K (2017) Toward a resource-efficient built environment: a literature review and conceptual model. J Ind Ecol 21:572–592

Norouzi M, Chàfer M, Cabeza LF, Jiménez L, Boer D (2021) Circular economy in the building and construction sector: a scientific evolution analysis. J Build Eng 44:102704

Nußholz JLK, Milios L (2017) Applying circular economy principles to building materials: front-running companies' business model innovation in the value chain for buildings. https://www.researchgate.net/publication/320831772. Accessed November 2022

OECD (Organisation for Economic Co-operation and Development) (2020) Environment at a glance indicators. Circular economy, waste and materials. https://www.oecd.org/environment/environment-at-a-glance/Circular-Economy-Waste-Materials-Archive-February-2020.pdf. Accessed December 2022

Ogunmakinde OE, Egbelakin T, Sher W (2022) Contributions of the circular economy to the UN sustainable development goals through sustainable construction. Resour Conserv Recycl 178:106023

PEEB (Programme for Energy Efficiency in Buildings) and GlobalABC (Global Alliance for Building Construction) (2022) The EU taxonomy – what does it mean for buildings? Briefing paper developed by PEEB. https://www.peeb.build/imglib/downloads/PEEB_EU_Taxonomy.pdf

Preston F (2012) A global redesign? Shaping the circular economy. Briefing Paper, Chatham House, London

Rahla KM, Mateus R, Bragança L (2019) Comparative sustainability assessment of binary blended concretes using supplementary cementitious materials (SCMs) and ordinary Portland cement (OPC). J Clean Prod 220:445–459

Rahla KM, Mateus R, Bragança L (2021) Implementing circular economy strategies in buildings. From theory to practice. Appl Syst Innov 4(2):26

Rios FC, Grau D (2020) Circular economy in the built environment: designing, deconstructing, and leasing reusable products. Encycl Renew Sustain Mater 5:338–343

Sáez-de-Guinoa A, Zambrana-Vasquez D, Fernández V, Bartolomé C (2022) Circular economy in the European construction sector: a review of strategies for implementation in building renovation. Energies 15(13):4747

Sauvé S, Bernard S, Sloan P (2016) Environmental sciences, sustainable development and circular economy: alternative concepts for trans-disciplinary research. Environ Dev 17:48–56

Terzioğlu N (2021) Repair motivation and barriers model: investigating user perspectives related to product repair towards a circular economy. J Clean Prod 289

Treiblmaier H (2018) The impact of the blockchain on the supply chain: a theory-based research framework and a call for action. Supply Chain Manag Int J 23(6):545–559

UN (United Nations) (2015) Transforming our world: the 2030 agenda for sustainable development. Resolution adopted by the general assembly on 25 September 2015. https://documents-dds-ny.un.org/doc/UNDOC/GEN/N15/291/89/PDF/N1529189.pdf?OpenElement. Accessed December 2022

Van Buren N, Demmers M, Van der Heijden R, Witlox F (2016) Towards a circular economy: the role of dutch logistics industries and governments. Sustainanility 8(7)

Standards and Laws

BS 8887-2:2009 Design for manufacture, assembly, disassembly and end-of-life processing (MADE). Terms and definitions

COM (2008) 400. Communication from the Commission to the European Parliament, the Council, the European Economic and Social Committee and the Committee of the Regions. Public procurement for a better environment. https://eur-lex.europa.eu/LexUriServ/LexUriServ.do?uri=COM:2008:0400:FIN:EN:PDF

COM (2015) 614. Communication from the Commission to the European Parliament, the Council, the European Economic and Social Committee and the Committee of the Regions. Closing the loop—An EU action plan for the Circular Economy. https://eur-lex.europa.eu/resource.html?uri=cellar:8a8ef5e8-99a0-11e5-b3b7-01aa75ed71a1.0012.02/DOC_1&format=PDF

Directive 2008/98/EC. Directive 2008/98/EC of the European Parliament and of the Council of 19 November 2008 on waste and repealing certain directives. https://eur-lex.europa.eu/legal-content/EN/TXT/?uri=celex:32008L0098

Directive 2009/125/EC of the European Parliament and of the Council of 21 October 2009 establishing a framework for the setting of ecodesign requirements for energy-related products. https://eur-lex.europa.eu/legal-content/EN/TXT/PDF/?uri=CELEX:32009L0125&from=EN

ISO/IEC 2382-14:1997 Information technology. Vocabulary. Part 14: Reliability, maintainability and availability

ISO/IEC 25010:2011 Systems and software engineering. Systems and software quality requirements and evaluation (SQuaRE). System and software quality models

Regulation (EC) No 66/2010 of the European Parliament and of the Council of 25 November 2009 on the EU Ecolabel. https://eur-lex.europa.eu/legal-content/EN/TXT/PDF/?uri=CELEX:32010R0066&from=EN

Regulation (EU) 2020/852 of the European Parliament and of the Council of 18 June 2020 on the establishment of a framework to facilitate sustainable investment, and amending regulation (EU) 2019/2088. https://eur-lex.europa.eu/legal-content/EN/TXT/PDF/?uri=CELEX:32020R0852&from=EN

Regulation EU 2020/852 Annex 1. Annex supplementing Regulation (EU) 2020/852 of the European Parliament and of the Council by establishing the technical screening criteria for determining the conditions under which an economic activity qualifies as contributing substantially to climate change mitigation or climate change adaptation and for determining whether that economic activity causes no significant harm to any of the other environmental objectives. https://ec.europa.eu/finance/docs/level-2-measures/taxonomy-regulation-delegated-act-2021-2800-annex-1_en.pdf

Websites

UN website. https://sdgs.un.org/

Chapter 2
Circularity Integration in Building Practices: Windows of Opportunity

Check for updates

2.1 Circularity Within Building Process: The Three Key Windows of Opportunity

The present paragraph aims to identify the "windows of opportunity" of the Building Process, meant as favorable moments for introducing reflections on the issues of environmental sustainability and, in particular, circularity. In the last decades, the concept of "window of opportunity"—namely an advantageous period of time during which decisions can be taken and actions can be implemented to solve a problem and to achieve a desired outcome—has been widely employed in literature with applications at different scales and in different fields, including technology and sustainability (Perez and Soete 1988; Tyre and Orlikowski 1994; Geels 2011; Doeser and Eidenfalk 2013; Van Stigt et al. 2013; Cairney and Jones 2016; Lusk 2016; Fuchs 2017; Lee and Malerba 2017; Borsboom-van Beurden 2018; Cooper-Searle 2018; Hernandez et al. 2018; Bos 2020; Win 2021; Derwort et al. 2022; Garrido et al. 2023; Von Malmborg et al. 2023), also in the context of the Multiple Streams Framework (MSF) (Kingdon 1984, 1993, 2001). The latter, developed by JW Kingdon (Kingdon 1984, 1993, 2001; Kingdon and Thurber 2010) in 1984 (then updated in 2010), can be defined as a methodological approach to analyze and handle decision-making processes. In particular, according to the MSF by Kingdon (Kingdon 1984, 1993, 2001; Kingdon and Thurber 2010), when the "problem, policy and politics streams" overlap a "window of opportunity" or a "window for change" arises. In particular, in the MSF (Fig. 2.1):

- the problem stream is represented by the evolution from a "situation" to a "problem". Indeed, not always a "situation" is definable as a "problem" but it depends from the perception of the situation and its consequences by a person or groups of people. In particular, a situation becomes a problem worthy of public attention—thus to be put on the "government agenda for action" in search of a solution (Kingdon 2001; Kingdon and Thurber 2010; Hoefer 2022)—when for instance:

N. Atta, *Green Approaches in Building Design and Management Practices*,
Digital Innovations in Architecture, Engineering and Construction,
https://doi.org/10.1007/978-3-031-46760-8_2

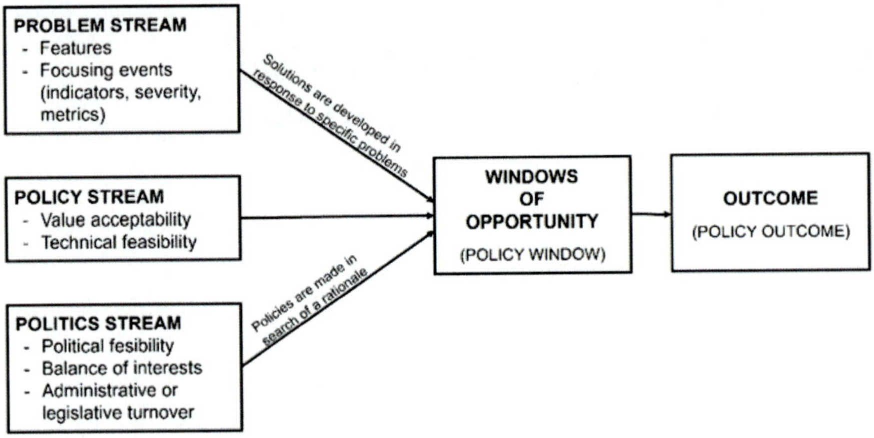

Fig. 2.1 The Kingdon's Multiple Streams Framework (MSF). *Source* Adapted from Mu (2018) and Jarvis et al. (2019)

dramatic events occur, the magnitude of the consequences of the problem is very high, the problem is shared by large masses, etc. (Kingdon 2001; Kingdon and Thurber 2010; Hoefer 2022);

– the policy stream represents the process of "solution development" by experts after the problem examination and analysis. Thus, the outcome of this stream is represented by the availability of a problem solution—highlighted and promoted by policies—characterized by value acceptability and technical feasibility (Mu 2018);
– the politics stream is defined as a combination of "*national mood, executive or legislative turnover and interest group advocacy campaigns*" (Béland and Howlett 2016) that converge when an opportunity for action emerges, namely when a defined problem is identified together with a feasible and acceptable solution (Kingdon 2001; Hoefer 2022).

The overlapping of the three streams generates a resulting time interval, called by Kingdon "policy window" (Kingdon 2001; Kingdon and Thurber 2010; Hoefer 2022), which represents—on the one hand—the opportunity to enrich the regulatory framework, promoting new normative tools (communications, regulations, laws, standards, etc.), pushing the majority of decision-makers and key actors to support and follow the new policy package (Hoefer 2022) and—on the other hand—the opportunity to push stakeholders to pursue and develop feasible added-value solutions to the identified problem towards the optimization and innovation of current practices (problem resolution).

In light of these premises, by adopting the MSF vision and the concept of "Window of Opportunity"[1] as starting points and guiding approaches to analyze the Building Process, it is possible to identify some key possibilities[2] to integrate circular principles within current building practices. In this regard, the scope of analysis is the whole Building Process (pre-design and design, construction, use and management, end-of-life phases) and the resulting three key "windows of opportunity" concern:

a. the *development of the Briefing Documents (BDs)* and the definition of project objectives and design principles (Sect. 2.2.1);
b. the *development of Invitations to Tenders (ITTs)* for the provision of building products and services related to Operation and Maintenance (O&M) (Sect. 2.2.2);
c. the *development of sustainability and circularity assessment reports* during the building management phase to adjust and/or improve strategies and practices according to a continuous improvement approach (Sect. 2.2.3).

2.2 Formalization of Circular Practices by Reviewing and Integrating Existing Documents and Tools

In relation to the previously introduced windows of opportunity, the present paragraph discusses the primary importance of the revision, updating and integration of traditional documental systems and information support tools, in accordance with the current regulatory framework on Circular Economy issues (Table 2.1). In particular, the following sub-paragraphs focus on the three windows of opportunity and explore the related key documental systems (process outcomes)—namely (a) Briefing Documents (BDs) for building design; (b) Invitation to Tenders (ITT) for products and services provision; (c) Sustainability and circularity assessment reports—opening to possible integrations of circular concepts and requirements (further detailed in Chaps. 3, 4 and 5).

---

[1] With reference to the building sector, in the face of the growing awareness of the criticality of circularity and its role in pursuing more sustainable practices and buildings (*problem stream*), also supported by the current regulatory framework (including EU taxonomy, Corporate Sustainability Reporting Directive, Level/s, etc.) (*politics stream*), the key stakeholders of the building process (including clients, project managers, design firms, quality and environmental sustainability consultants, manufacturers, construction companies, facility managers, users, etc.) can co-operate in synergy to gradually develop new feasible approaches and added-value strategies towards circularity (*policy stream*) by optimizing current practices according to an incremental innovation logic. In light of this assumption, the overlapping of the three streams opens "windows of opportunity" across the Building Process, meant as favorable moments of the building process in which it is possible to integrate current practices with circular approaches oriented towards the reuse and remanufacturing of building elements (products/components) in order to lengthen their useful life and maximize their use cycles (see the circular strategies and models introduced in Chap. 1).

[2] Defined as those "useful moments" to integrate circularity-oriented concepts, methods, tools and practices, thus also integrating the related decision-making processes with new dimensions related to circular economy and circularity.

Table 2.1 Key voluntary and mandatory regulatory references on the topics of Circular Economy and circularity at the European level (non-exhaustive list)

EU communications	
COM/2014/398	Communication from the Commission to the European Parliament, the Council, the European Economic and Social Committee and the Committee of the Regions "*Towards a circular economy: A zero waste programme for Europe*"
COM/2019/640	Communication from the Commission to the European Parliament, the Council, the European Economic and Social Committee and the Committee of the Regions "*The European Green Deal*"
COM/2020/98	Communication from the Commission to the European Parliament, the Council, the European Economic and Social Committee and the Committee of the Regions "*A new Circular Economy Action Plan For a cleaner and more competitive Europe*"
COM/2020/408	Proposal for a Regulation of the European Parliament and of the Council establishing a *Recovery and Resilience Facility*
COM/2021/573	Communication from the Commission to the European Parliament, the Council, the European Economic and Social Committee and the Committee of the Regions "*New European Bauhaus Beautiful, Sustainable, Together*"
EU regulations and directives	
2008/98/EC Directive (Waste Directive)	Directive 2008/98/EC of the European Parliament and of the Council of 19 November 2008 on waste and repealing certain Directives
2014/24/EU Directive (Green Public Procurement)	Directive 2014/24/EU of the European Parliament and of the Council of 26 February 2014 on public procurement and repealing Directive 2004/18/EC
2018/851 Directive (EPR–Extended Producer Responsibility)	Directive (EU) 2018/851 of the European Parliament and of the Council of 30 May 2018 amending Directive 2008/98/EC on waste
2019/2088 Regulation (SFDR—Sustainable Finance Disclosure Regulation)	Regulation (EU) 2019/2088 of the European Parliament and of the Council of 27 November 2019 on sustainability-related disclosures in the financial services sector
2020/852 Regulation (EU Taxonomy)	Regulation (EU) 2020/852 of the European Parliament and of the Council of 18 June 2020 on the establishment of a framework to facilitate sustainable investment, and amending Regulation (EU) 2019/2088
2022/2464 Directive (CSRD—Corporate Sustainability Reporting Directive)	Directive (EU) 2022/2464 of the European Parliament and of the Council of 14 December 2022 amending Regulation (EU) No 537/2014, Directive 2004/109/EC, Directive 2006/43/EC and Directive 2013/34/EU, as regards corporate sustainability reporting

(continued)

Table 2.1 (continued)

2022/1288 Regulation	Commission Delegated Regulation (EU) 2022/1288 of 6 April 2022 supplementing Regulation (EU) 2019/2088 of the European Parliament and of the Council with regard to regulatory technical standards specifying the details of the content and presentation of the information in relation to the principle of 'do no significant harm', specifying the content, methodologies and presentation of information in relation to sustainability indicators and adverse sustainability impacts, and the content and presentation of the information in relation to the promotion of environmental or social characteristics and sustainable investment objectives in precontractual documents, on websites and in periodic reports
Standards and guidelines at the European level	
BS 8887-1: 2006	BS 8887: 2006 Design for manufacture, assembly, disassembly and end-of-life pro-cessing (MADE)—Part 1: General concepts, process and requirements
BS 8887-2: 2009	BS 8887: 2009 Design for manufacture, assembly, disassembly and end-of-life processing (MADE)—Part 2: Terms and definitions
BS 8887-3: 2018	BS 8887: 2018 Design for manufacture, assembly, disassembly and end-of-life processing (MADE)—Part 3: Guide to choosing an appropriate end-of-life design strategy
BS 8887-220: 2010	BS 8887: 2010 Design for manufacture, assembly, disassembly and end-of-life processing (MADE)—220: The process of remanufacture. Specification
BS 8887-240: 2011	BS 8887: 2011 Design for manufacture, assembly, disassembly and end-of-life processing (MADE)—240: Reconditioning
ISO 20887: 2020	ISO 20887: 2020 Sustainability in buildings and civil engineering works. Design for disassembly and adaptability. Principles, requirements and guidance
ISO/DIS 59020: 2023	ISO/DIS 59020: 2023 Circular economy. Measuring and assessing circularity
Level(s)—European framework for sustainable buildings (2019—in progress)	Level(s) European framework for sustainable buildings (2019), i.e.: 1.2 Life cycle global warming potential (CO_2 eq./m^2/yr) 2.1 Bill of quantities, mate-rials and lifespans 2.2 Construction and demolition waste (CDW) and materials 2.3 Design for adaptability and renovation 2.4 Design for deconstruction, reuse and recycling 6.1 Life cycle costs (€/m^2/yr) 6.2 Value creation and risk factors

2.2.1 Development of the Briefing Documents (BDs) for Building Design. Definition of Project Objectives and Design Principles

The definition of the project objectives and design principles represents the starting point of the Building Process. In particular:

- project objectives clarify the intents of the project, i.e. what needs to be accomplished through and with the building project. They guide the project design stage, providing purpose and direction to the subsequent design activities. The project objectives are of a general nature and they do not refer to the building systems and parts but they focus on the value that the building project as a whole has to deliver to the Client and end-users;
- design principles represent the overall concepts, general aims or high-level requirements that the design project has to follow and pursue. They can refer to the building as a whole or they can focus on one or more aspects or building parts or systems. They can include overall requirements regarding the broad architectural expression as well as concerning specific topics (e.g. health and wellbeing, sustainability, resource efficiency, operation and maintenance, space flexibility, accessibility, etc.).

The project objectives and design principles are stated, described and further detailed in the form of functional and technical requirements within the Briefing Documents (or Client's Brief) (Fig. 2.2), a key preliminary documentation that represents a precondition to set the stage for the subsequent building design phase.

The Briefing Documents (BDs) represent the outcome of the gradual and reiterative process of defining the client requirements for the development of a building. In particular, the drafting of the final Briefing Documents (BDs) often implies the progressive development of a series of preparatory documents, e.g.:

i. the "statement of needs", that is a first attempt to explicit the broad client needs into macro-functions of the building project;
ii. the "strategic brief", that translates the contents of the "statement of needs" into elementary functions and, then, measurable/assessable requirements of the building project;
iii. the "project brief", that lastly better clarifies the project requirements and any related acceptability and tolerance thresholds on the basis of the information collected by the design team through consultations with the client and other key involved stakeholders and experts.

This process requires the collaboration among different stakeholders (design teams, client, end-users, environmental consultants, etc.) and it represents an opportunity to include and share circularity objectives for the building project and to define the related circular design principles.

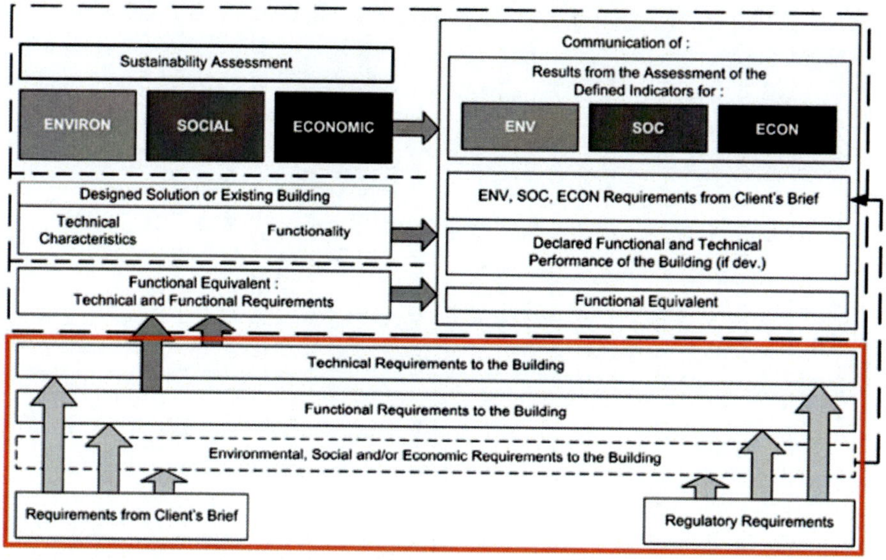

Fig. 2.2 The contribution of the project brief document to the definition and pursuit of project requirements. *Source* (Adapted from) BS EN 15978:2011

Hence, new circularity principles and strategies (Table 2.2)—oriented to design-for-disassembly, design-for-remanufacturing, extended-use of products, closed-loop models for resource management, etc.—must be taken into account when defining the building design principles in order to lead the whole project towards Circular Economy objectives.

As shown in Table 2.3, the three CE principles introduced in Table 2.2 can be further detailed according to the different addressed "Target Groups" (EU 2020), i.e.: (i) Building users, facility managers and owners; (ii) Design teams; (iii) Contractors and builders; (iv) Manufacturers of construction products; (v) Deconstruction and demolition team; (vi) Investors, developers and insurance providers.

Hence, adopting circular approaches in this very first preliminary stage of the Building Process is fundamental to anticipate some key lifecycle considerations, including the possible future re-strategies (reuse, remanufacturing, repurposing, etc.) to implement for life-extension of building elements (de Graaf et al. 2022). Indeed, most of the footprint impact of buildings takes place during the use phase. Circular strategies should be integrated at the beginning of the design phase in order to guide the design towards a sustainable construction in terms of employed materials and products but also in terms of possibility to use and manage the building during its whole life in an effective and efficient way (e.g. energy consumptions, maintenance interventions, circular manufacturing, resource management, new product-as-a-service business models, product reuse-ability, digital innovations, lifetime extension, product ease to repair, disassembly or remanufacture, etc.), retaining the value embedded in the building overtime and closing the loop of products by maximizing

Table 2.2 Examples of Circular Economy principles for conceptual design. *Source* Adapted from EU (2020)

CE principle	Aim and description	Strategy	Actions and requirements for the project brief
Durability	Building and elements service life planning, encouraging a medium to long term focus on the design life of major building elements, as well as their associated maintenance and replacement cycles	Durability of buildings depends on the performance of building products and systems and their expected useful life. Building elements should last as long as the building does, whenever possible. If it is not possible because of intrinsic obsolescence or anticipated change in requirements, they should be reusable, recyclable or dismountable	– Favour construction systems that incorporate circular economy thinking. For instance, enable systems to be easily maintained, repaired and replaced as this will extend the life cycle of buildings – Require detailed information from providers and designers on products, materials and the design of the buildings. Conserve, update and share the information so that it can remain valid and relevant during the whole life cycle of the building
Adaptability	Extend the service life of the building as a whole, either by facilitating the continuation of the intended use or through possible future changes in use (with a focus on replacement and refurbishment)	Prevent premature building demolishment by developing a new design culture	– Anticipate changes in requirements – Enable adaptations and transformations of the building and its parts for a better future use and reuse (as well as new possible ways of using the building components) – Require the collection of information useful to better manage the maintenance, end-of-life and future lives of the building and its components

(continued)

Table 2.2 (continued)

CE principle	Aim and description	Strategy	Actions and requirements for the project brief
Reduce waste and facilitate high-quality waste management	Facilitate the future circular use of building elements, components and parts, with a focus on producing less waste and on the potential for the reuse, or high-quality recycling, of major building elements following deconstruction. This includes efforts along the value chain to promote: – The reuse or recycling of resources (i.e. materials) in a way that most of the material value is retained and recovered at the end of a building life span – Design-for-dis/assembly (DfD/A) and proper construction methods to allow the recovery of products for reuse, remanufacturing and recycling	Design products and systems so that they can be easily disassembled, reused, repaired, recycled or recovered	– Define design-for-dis/assembly (DfD/A) strategies to employ easy-to-dismount elements and products – Require the use of simple and reusable/recyclable products

Table 2.3 Examples of Circular Economy principles translated into specifications for the different addressed "Target Groups". *Source* Adapted from EU (2020)

	Target group	Key focus objectives	Specifications for the target group of stakeholders
Durability	Design teams (engineering and architecture of buildings)	Gain knowledge of circular economy principles to design buildings and materials. Architects and designers should be familiar with the circular design requirements and strategies, the concept of life cycle assessment, the potential to increase the content of recycled materials in products, future reuse potential (product, component and building); (future) recyclability and transformation capacity (reuse potential and reversible building design score)	– Encourage architects and designers to adopt a life cycle approach when designing new buildings – Use regulatory references and existing guides on design-for-dis/assembly DfD/A as well as feedback from previous project examples
		Take into account whole life costs and benefits	The whole life cycle must take into account the operational cost of the building as well as the potential changes to the building use. They include environmental and social impacts and benefits, transformation capacity, reuse and recyclability potential
		Project management team needs to be engaged in the process and to consider assessment methods	Establish a set of relevant indicators with regard to the overall circular management system in a building, taking into account those provided by EU Level(s)
	Contractors and builders	Use construction techniques that promote the durability of buildings and the resilience of the materials	– Simulate different scenarios of durability and compare costs – Include the resources needed for resilience to installation error – Enhance the building durability and use construction techniques that facilitate maintenance of building products and systems

(continued)

Table 2.3 (continued)

	Target group	Key focus objectives	Specifications for the target group of stakeholders
	Manufacturers of construction products	Consider the potential durability level for the whole life cycle of the building (e.g. assessed on the basis of the evidence from life cycle costing—LCC of the products)	Use whole life costing and environmental assessment methods, considering the entire life of the building and its parts
		Eco-design principles should be adopted and durability assessed	Product standards should include durability and a verification system should be adopted to confirm such durability
		Solutions should be developed for greater adaptability	Adaptability can be reached preferring prefabrication and modular systems
	Building users, facility managers and owners	Minimise the total cost of ownership over time. Owners and building users have an interest in overall and longer-term horizons	– Reduce the total cost per square meter/comparative average – Use tools to enhance the building value
		Promote durability during the use phase	Provide incentives through performance-based contracts that promote the optimal use of the building
		Perform appropriate preventive maintenance of the building and its parts to minimise running costs. The life cycle of a building can be extended by maintaining and repairing the building while reducing the use of new resources and the production of waste; information and guidelines help achieve proper maintenance and use to meet these aims	– Develop a preliminary maintenance plan for the building and its parts – Consider innovative circular business models and elaborate guidelines concerning the elements of fittings (e.g. carpets, false ceilings, doors, vertical partitions, light terminals, etc.), in terms of their selection, installation, maintenance, and disposal
	Investors, developers and insurance providers	Enhancing durability will decrease financial risk	The importance of durability of products and materials should be promoted within the overall approach to buildings and products (including assessment on how this can be appropriately accounted financially)

(continued)

Table 2.3 (continued)

	Target group	Key focus objectives	Specifications for the target group of stakeholders
		Life Cycle Costing (LCC) should be promoted when preparing investment decisions. The increased revenue streams that can be generated through reversible design should be integrated into the whole cost analysis	– Capitalise future risks of difficulty to deconstruct buildings and cost of waste management – The use of the ISO standards for DfD/A (e.g. ISO 20887) credits within Green Public Procurement (GPP) and sustainable building rating schemes provide an incentive to consider at this stage
Adaptability	Design teams (engineering and architecture of buildings)	Promote and ensure reversibility and adaptability of the building (the periods between changes of use, renovation or reconstruction are becoming shorter)	Include in the pre-design stage preliminary analyses concerning aspects such as building resilience, reversibility and functional adaptability (e.g. identifying different use scenarios for the building and its parts)
		Assess the proposed building design for suitability	The "suitability" addesses materials, products and systems based on their technical characteristics, their adaptability and life span (technical and economical) and their environmental impact. This will facilitate the maintenance and repair of the building while reducing the use of new resources and the creation of waste
	Contractors and builders	Depending on the frequency of maintenance, repair and transformation needs, it is possible to define several stages of reversibility for various parts of the building, allowing for different construction techniques	Use construction techniques that promote the adaptability of buildings
	Building users, facility managers and owners	Minimise the financial cost of use through adequate tools. Such tools support ease of maintenance, adaptations, repair, monitoring and running costs. It is important to inform the users and facility managers of the reuse potential of the building and its parts	– BIM and building passports can support the transfer of information and apprise the users and maintenance team on how to best use/maintain/repair or adapt the building and its systems and products – A user guide for the building and its equipment is useful for owners/users of the building and its equipment

(continued)

Table 2.3 (continued)

	Target group	Key focus objectives	Specifications for the target group of stakeholders
		The building should be adapted if it extends the lifetime at reasonable cost. Adaptability can improve the response to shifts in market demand (e.g. transforming an unused office building into apartments). Adaptability can also help secure financing: for example, the potential of a second life cycle of a building can ensure cheaper financing and loan insurance	Assess the cost/benefit ratio of adaptability of the building and its parts as well as of the product re-strategies (remanufacturing, repurposing, etc.)
		Promote and maximise the adaptability of buildings from a user perspective. Information and guidelines should help users, owners, and facility managers to make modifications to the building and so have a lower impact on waste generation	– The possibility to reuse building materials when the use is changed should be assessed. – Apply reversible building design principles to support the flexible use and ease of adaptation of spaces (for when the purpose and requirements of a building will change)
	Investors, developers and insurance providers	Adaptability of buildings should be appropriately accounted for financially. Any measure that is of benefit to the owner will minimise the risk of default	Building owners should consider the importance of adaptability within the overall approach to buildings and products, and how this can be appropriately accounted for financially
Reduce waste and facilitate high-quality waste management	Design teams (engineering and architecture of buildings)	Use materials that are easy to recycle or reuse, and which facilitate high quality waste management	– Employ products with a high content of recyclable and reusable matter – Consult the environmental product declarations, product safety data sheets, LCAs, etc. – Enable designers to explore possibilities of using products and materials already available within existing buildings

(continued)

Table 2.3 (continued)

	Target group	Key focus objectives	Specifications for the target group of stakeholders
		Take into account the variety of circular economy aspects when designing the different systems and elements	Consider aspects such as: the size/volume/weight, etc., of materials to manage in the demolition process; functional decomposition; disassembly procedures; hierarchical relations between elements; product specifications; assembly sequences; geometry of the connections; type of connections; life cycle co-ordination in assembly/disassembly and the recyclability of materials and reusability of products, and how material choice can influence the quality of waste management
	Contractors and builders	For maximum benefits, construction, demolition or renovation need to be planned carefully and in advance. This can bring about important cost benefits, as well as environmental and health benefits and carbon savings	Carry out a pre-demolition audit (or waste management audit) before any renovation or demolition project to identify materials to be re-used or recycled (resources), as well as hazardous waste
		Facilitate deconstruction by using construction techniques as set out in design guidelines and standards	– Use construction techniques which contribute to easy and clean building deconstruction – Choose techniques/assembly systems of construction products that allow at least selective deconstruction (sorting at source) and at best dismantling – Support the uptake of BIM to facilitate the future building end-of-use/life activities, e.g. deconstruction, recycling and reuse
		Give preference to use of recycled, recyclable, reusable and/or reused products	Check the purchase documents, the products technical datasheet, and the availability of recycled and used matierials and products in the area

(continued)

Table 2.3 (continued)

	Target group	Key focus objectives	Specifications for the target group of stakeholders
	Manufacturers of construction products	Increase products recycling and recovery potential by providing the necessary information. Knowing what the product consist of is very important with regard to reuse potential and recycling, to avoid, for instance, contamination of an identified material stream with an unidentified one	Ensure good traceability of where/how the products are manufactured and collect technical datasheets from manufacturers
		Develop digital product passports according to the guidelines of the mandatory and voluntary regulatory framework. The product passports will support decision-making processes of designers and building managers during the design phase and the use phase of the building	– Know for what purpose the product has been designed (e.g. repair, reuse, remanufacturing, reconfiguration) – Know how the product is implemented/installed in, and connected to, the building and other construction products and systems – Keep available information about the technical characteristics of materials and products and promote traceability of the changes and uses of the product during its life cycle
		Minimise the use of natural resources of construction products wherever feasible. It is best to choose reused or recycled materials that provide durability, technical and environmental performance, and that meet the same maintenance requirements and standards of the primary material	– Provide information to the users and decision-makers on the reuse, recycling and recovery potential of construction products and materials to encourage a reduction of natural resource depletion – Use quality materials and with intrinsic finishing – Design by using standard dimensions so to reduce off-cuts

(continued)

Table 2.3 (continued)

	Target group	Key focus objectives	Specifications for the target group of stakeholders
		Avoid hazardous substances. This refers to substances of very high concern (SVHC) and other substances, solutions, and material compositions which could hamper reuse or recycling, or curb their use in public and private buildings, wherever feasible, as this will make future reuse/recycling more difficult	Alert the relevant actors about the presence of hazardous substances within products
	Building users, facility managers and owners	Minimise the use of natural resources during the building life. A reusable building parts inventory will inform the users and facility managers both during the life cycle of a building and in the next life cycles	– Establish contractual arrangements to return non-used items to providers – Develop a maintenance plan to extend the useful life of building products and systems
	Deconstruction and demolition team	Recycling/remanufacturing conditions must be defined and assessed. Circular value chains must be activated. Distances between the demolition site, the recycling/remanufacturing facilities, the (re)manufacturing plants and new construction site must be measured and evaluated (both in economic and environmental terms).	– Regional reuse, remanufacturing and/or recycling networks must be developed – Establish rules for the assessment of the sustainability of the regional loops (e.g. distances between sites, costs, environmental impacts, etc.)

(continued)

Table 2.3 (continued)

	Target group	Key focus objectives	Specifications for the target group of stakeholders
		Make effective use of pre-deconstruction or pre-development audit and appropriate information tools	– Identify the resources contained in the building and identify outlets and forms of recovery – Ensure selective deconstruction and take full account of state of the art practices – Assess construction and demolition materials and waste streams prior to deconstruction or renovation – Make use of tools developed (e.g. BIM, materials passports of construction products and systems as well as building passports) to enable rapid and accurate assessments of recovery, reuse and recycling potential of specific products and systems and corresponding value propositions
		Tailored dismantling techniques should be applied	The sorting must be made with dismantling; the materials must be separated into batches with a place of destination/treatment according to a "road map" (pre-demolition audit), as accurately as possible, established based on a list of products
		Preliminary sorting should take place on site. Machinery can support workers and improve safety, whilst new scanning techniques can help with material identification	– Assess the available space on site to perform the disassembly and sorting and, if needed, identify a sorting infrastructure nearby – Separate on site at least hazardous and non-hazardous waste – Perform disassembly according to the identified dismantling techniques (disassembly plan).
	Investors, developers and insurance providers	The use of recycled and recyclable materials should be appropriately accounted for financially. Finance/Insurance companies can set requirements for investing in projects	– Define circularity requirements for investing in building projects – Assess financially the benefits of circular strategies applied to the building project – Define standards for a circularity due diligence

their use and keeping them in use as long as possible (de Graaf et al. 2022). In particular, according to de Graaf et al. (2022), it is possible to identify four main goals related to the material loops that have to be considered in the design process in order to guide the choice of products and related procurement and business models, i.e. (Fig. 2.3):

1. "*Narrowing resource loops (reducing the input of resources) by refusing the use of products (prevention) when possible, intensifying the use of products or reducing the use of materials through more efficient manufacturing or efficiency in using them;*
2. *Slowing down or elongating resource loops (longer and high value use of materials and products) by reuse, repair, and remanufacturing of products;*
3. *Closing the loops (reducing loss of materials through waste) by recycling and recovering energy from materials when all the previous options are no longer possible;*
4. *Substitution where applicable. This includes the use of biobased, renewable materials instead of primary abiotic materials*" (de Graaf et al. 2022).

In light of these premises, the contents of the Briefing Documents (BDs) should be updated and integrated with the new circularity specifications and requirements (Tables 2.2 and 2.3) as result of a multi-stakeholder process of assessing the feasibility of the adoption of circular approaches in building design, use and management.

This topic will be deepened in Chap. 3 of the present book.

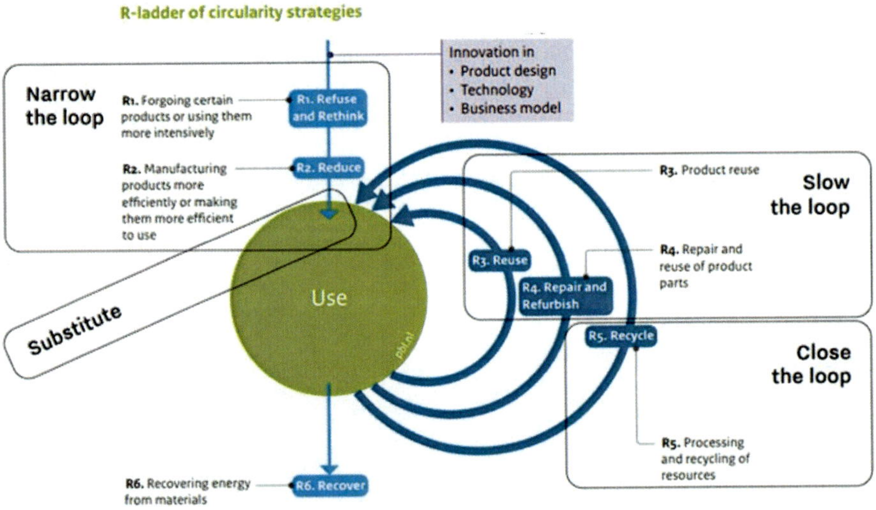

Fig. 2.3 The four main goals of the material loops according to Potting et al. (2018). *Source* de Graaf et al. 2022

2.2.2 Development of Invitations to Tenders (ITTs) for the Provision of Building Products and Services

The procurement of goods and services in the building use and management phase represents a key process to integrate policies, strategies and actions towards the circular management of the building during it whole life-cycle. This long-lasting phase, especially with reference to public and/or complex buildings, involves:

- the periodic substitutions of components, equipment and furniture, especially in the case of tertiary buildings, i.e. commercial, office and exhibition buildings;
- the regular maintenance of building elements with the related supply of spare parts and products;
- the integrated management of the no-core services, i.e. Facility Management (FM) services to "Space & Infrastructure" and "People & Organization" detailed in Table 2.4 (BS EN 15221-1: 2006);
- the purchase of products (e.g. consumables, cleaning products, etc.,) useful to the ordinary running of the building;
- etc.

All the points above introduced require a procurement process of products and/or services. To support stakeholders in the decision-making related to the procurement process, orienting their choices towards sustainability, the EU published in 2008 the COM 2008/400 "Public procurement for a better environment" that introduces the concept of Green Public Procurement (GPP). In particular, the COM 2008/400 defines the GPP as a "*process whereby public authorities seek to procure goods, services and works with a reduced environmental impact throughout their life cycle when compared to goods, services and works with the same primary function that would otherwise be procured*" (COM 2008/400). By integrating sustainability and circularity principles within the traditional goods/services procurement processes, the GPP promotes closed energy and material loops within supply chains, minimizing negative environmental impacts and waste creation during the building lifecycle (EU 2023a).

More recently, the EU Action Plan for the Circular Economy (APCE) introduced the concept of "Circular Procurement" as a key driver in the transition towards the circular economy (EU 2023a) and in the achievement of the Sustainable Development Goals (SGDs), with particular reference to Goal 12—Responsible Consumption and Production. According to EU, the implementation of Circular Procurement, i.e. integrating Circular Economy principles within traditional procurement practices, must follow a gradual approach that involves three different levels EU (2023a), namely: System Level, Supplier Level and Product Level (Fig. 2.4).

Table 2.4 Facility Management (FM) services to "Space & Infrastructure" and "People & Organisation" according to BS EN 15221. *Source* Adapted from BS EN 15221-1:2006

Service area	Description	Service and service activities
Services to Space & Infrastructure		
Accommodation	The client demand for space (accommodation) is satisfied by services such as programming, design and acquisition of space, but also the administration and management of space and its disposal	Strategic space planning and management
		Programming and briefing
		Design and construction
		Lease and occupancy management
		Building operations and maintenance
		Renovation and/or refurbishment
Workplace	The client demand for a working environment (workplace) is satisfied by services related to internal and external environments, fitting out with furniture, equipment and tenants	Workplace design and ergonomics
		Selection of furniture, machinery and equipment
		Move management
		Equip internal and external environment
		Signage, decorations, partitions and furniture replacement
Technical infrastructure	The client demand for utilities (technical infrastructure) is satisfied by services resulting in a comfortable climate, lighting/shading, electrical power, water and gas	Energy/utilities management
		Environmental sustainability management
		Technical infrastructure operations and maintenance
		Building management systems operations and maintenance
		Lighting maintenance
		Management of waste (hazardous) disposal
Cleaning	The client demand for hygiene and cleanliness (cleaning) is satisfied by services that maintain a proper working environment and help maintain the assets in good condition	Hygiene services
		Workplace cleaning, machinery cleaning
		Building fabric and glass cleaning
		Cleaning equipment provision and maintenance
		Outdoor space cleaning and winter services
Other space and infrastructure	Specific or individual demands of clients related to space and infrastructure	Hiring of special measuring equipment
		Fitting out with machinery and equipment
		Retail unit space management

(continued)

Table 2.4 (continued)

Service area	Description	Service and service activities
Services to People & Organization		
Health, safety and security	The client demand for a safe environment (health, safety and security) is satisfied by services that protect from external dangers or internal risks as well as the health and well-being of the people	Occupational health services
		Security management
		Access control, ID/smart cards, locks and key holding
		Disaster planning and recovery
		Fire safety and protection
Hospitality	The client demand for hospitality is satisfied by services providing a hospitable working environment makes people feel welcome and comfortable	Secretarial and reception services
		Help desk services
		Catering and vending
		Organisation of conferences, meetings and special events
		Personal services
		Provision of work wear
ICT	The client demand for Information and Communication Technology (ICT) is satisfied by services providing information and telecommunication technologies	Data and telephone network operations
		Data centre, server hosting and operations
		Personal computer support
		IT security and protection
		Computer and telephone connections and moves
Logistics	The client demand for logistics is satisfied by services concerned with the transport and storage of goods and information and improving the relevant processes	Internal mail and courier services
		Document management and archiving
		Reprographic systems, copying and printing
		Office supplies
		Freight forwarding, storage systems
		People transport and travel services
		Car park and vehicle fleet management
Other support services	The client demand for other support services may be satisfied by a range of additional personalised services (that may vary also depending on the primary activities of the client)	Accounting, auditing and financial reporting
		Human resource management
		Marketing and advertising, photographic services
		Procurement, contract management and legal advice services
		Project management
		Quality management

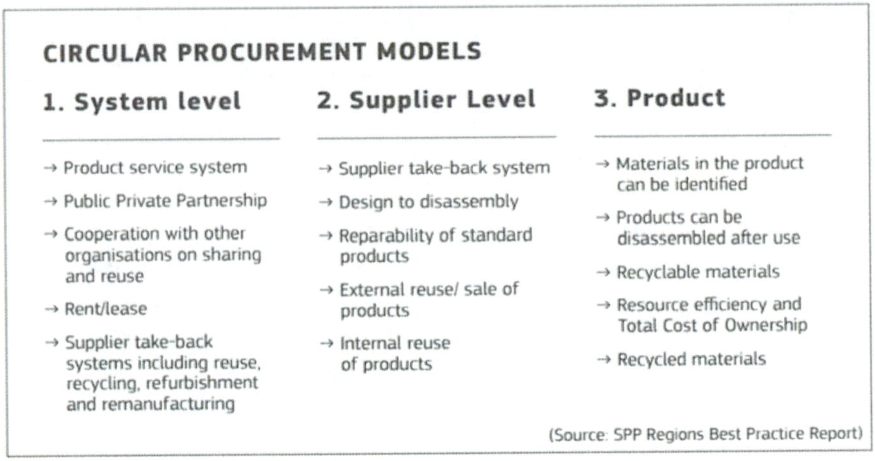

Fig. 2.4 The three levels of the Circular Procurement. *Source* EU (2023a)

In particular:

- the System Level refers to circular business models (e.g. renting, leasing, as-a-service, product-plus-services, deposit-based, etc.), including new goods/services selling modalities, as well as related contracts/agreements and payment methods (pay-per-use, pay-per-period, pay-per-performance, etc.);
- the Supplier Level concerns the circularity-related process specifications (e.g. for service delivery process, etc.) to suppliers/providers aimed at ensuring that the products and services they offer meet the circular procurement principles and criteria;
- the Product Level involves the product requests oriented towards circularity (e.g. design-for-reuse/remanufacturing specifications, including ease-to-disassembly), technical specifications (including expected useful life, durability, maintainability, etc.) and product certifications (including EPD—Environmental Product Declaration, LCA—Life Cycle Assessment, etc.).

The implementation process requires efforts and clarity of purposes. Therefore, it is necessary to have a clear understanding of the "Procurement Hierarchy" (EU 2023a), which is based on the EU Waste Hierarchy: "prevention, preparing for reuse reduce, recycling, recovery, disposal (2008/98/EC Directive)" in order to be able to define which closed-loop contracts and circular strategies (Table 2.5) are actually implementable and how to define the related requests in the ITT documents.

Table 2.5 Circular strategies of the Procurement Hierarchy in the context of circular procurements. *Source* Adapted from EU (2023a)

New type of contract	Description	Circular strategy to describe in ITTs
Product—service contract	The provider retains the ownership of the product and the client pays for the use of the product for a limited timeframe through pay-per-use, pay-per-period or pay-per-performance formulas	Reduce
		Reuse
Purchase and buy-back contract	The provider buys back a product and ensures optimum value retention via reuse or remanufacturing	Reduce
		Reuse
		Remanufacture
Purchase and resale contract	The contract includes an agreement on who (the provider or a third party) will recover the product after use, normally for lower-value material reuse, repurposing (new function) or recycling (downgrading)	Reduce
		Repurpose
		Recycle

In particular, in the context of ITTs:

- *reduce* is achieved by avoiding buying new products but preferring to buy a set of services for the life-extension of products. Moreover, reduction can also be achieved by opting for products or spare parts that have no packaging or by replacing just a component instead of the whole product or system when possible;
- *reuse, remanufacture and repurpose* involve the definition of specifications within ITTs focused on the propension of the product to be used for multiple subsequent cycles. Together with the requests concerning durability, ease to dis/assembly, etc. it is important to include contractual clauses aimed at the take-back of products that oblige or incentivize (deposit-based) the provider to take back the product at the end of the first use to start a second cycle of use and then a subsequent one;
- *recycle* can be meant in the ITT in two ways, i.e. content of recycled matter and content of recyclable matter of a product (resource efficiency). The client must specify its needs with respect to recycled and recyclable content of products. It is intended that recycling is a process not to prefer to reuse;
- *recover* is the process to convert waste into raw matter for different purposes. The recovery can be included into ITTs in two ways, i.e. (i) the request for strategies of design-for-recovery to define from the beginning the end-of-useful lives of post-use products (with insufficient residual performance) and (ii) the procurement of recovered products, hence products produced from waste/scraps. Recovery can also involve industrial symbiosis approaches.

Hence, the ITT represents a key document in which the client can clearly define and share with potential providers its circularity intents and purposes in order to ensure the subsequent success of the service/good delivery. In this way, the client can encourage the provider to be equally responsible towards the pursuit of sustainability and circularity objectives (e.g. extended responsibility, including the request to the provider to take responsibility for keeping post-use products in the supply chain, provision of support maintenance service package, etc.).

This topic will be deepened in Chap. 4 of the present book.

2.2.3 Development of Sustainability and Circularity Assessment Reports

Sustainability assessments during the building management phase represent recognized valuable tools to adjust and/or improve strategies and practices of building management beside being mandatory for European large companies. In particular, a regulatory framework on this subject at the European level is already present (Non-Financial Reporting Directive—NFRD[3]) and it has been recently updated (in 2022) with the new Corporate Sustainability Reporting Directive (CSRD).[4] On 5 January 2023, the CSRD entered into force. This new Directive reviews and reinforces the rules concerning the social and environmental sustainability report—and the required data and information to include—that companies have to draft and publish annually.

In particular, the CSRD applies to (EY 2022):

- Large companies. Large companies include companies that meet two of the following three conditions: (a) a net turnover of €40 million; (b) a balance sheet total of €20 million; (c) 250 on average over the financial year. In addition, the CSRD also applies to the non-EU companies with substantial business activity in the EU market that have a turnover of above €150 million in the EU and which have at least one subsidiary (large or listed) or branch (with a net turnover of more than €40 million) in the EU.
- Small and medium enterprises (SMEs). The CSRD applies to small companies with securities listed on regulated markets. However, they can report using simplified standards. Moreover, the European Commission has also proposed separate simplified standards for non-listed SMEs (standards are properly scaled to fit SMEs capabilities) that they could voluntarily use.

In particular, the European Commission adopted the CSRD in 2022 and the related rules will start to be applied between 2024 and 2028 as follows (EY 2022):

- From 1 January 2024 for large public-interest companies (with over 500 employees) already subject to the Non-Financial Reporting Directive (NFRD), with reports due in 2025 on 2024 data;

[3] Directive 2014/95/EU of the European Parliament and of the Council of 22 October 2014 amending Directive 2013/34/EU as regards disclosure of non-financial and diversity information by certain large undertakings and groups.

[4] Directive (EU) 2022/2464 of the European Parliament and of the Council of 14 December 2022 amending Regulation (EU) No 537/2014, Directive 2004/109/EC, Directive 2006/43/EC and Directive 2013/34/EU, as regards corporate sustainability reporting.

- From 1 January 2025 for large companies that are not subject to the NFRD (with more than 250 employees and/or €40 million of turnover and/or a balance sheet total of €20 million in total assets), with reports due in 2026 on 2025 data;
- From 1 January 2026 for listed SMEs, with reports due in 2027 on 2026 data. SMEs also have an opt-out option until 2028 to report.
- From 1 January 2028 for third-country companies, with reports due in 2029 on 2028 data.

For all these companies—approximately 50,000 companies in total (European Parliament 2022; EU 2023b)—the drafting of the Sustainability Report can be an opportunity to integrate and assess circular processes and closed-loop models. An effort on this side has been done by the EU with:

- the European Sustainability Reporting Standards (ESRS). According to the CSRD, the Sustainability Report should be drafted according to the European Sustainability Reporting Standards (ESRS). The draft standards are developed by the EFRAG, previously known as the European Financial Reporting Advisory Group. The ESRS deepen in terms of requirements three different key areas of interest (topical standards) concerning Environmental, Social and Governance (ESG) aspects. Among the Environmental Standards (E) which include for instance climate change, pollution, water and marine resources, biodiversity and ecosystems, there are the disclosure requirements concerning circularity, namely: "E5 Resource use and circular economy". In particular, the Environmental Standard E5 focuses on: policies (E5-1), actions (E5-2) and target (E5-3) related to resource use and circular economy; resource inflows (E5-4) and outflows (E5-5), potential financial effects from resource use and circular economy-related impacts, risks and opportunities (E5-6) (EFRAG 2022).
- The EU Taxonomy (Taxonomy Regulation—2020/852/EU), published in 2020, defines a list of environmentally sustainable activities by defining technical screening criteria for each of its six environmental objectives. In particular, among the six climate and environmental objectives established by the EU Taxonomy Regulation (i.e. climate change mitigation, climate change adaptation, sustainable use and protection of water and marine resources, pollution prevention and control, protection and restoration of biodiversity and ecosystems) there is the objective "Transition to a circular economy" that specifies the circularity requirements for the construction of new buildings and the renovation of existing buildings. With reference to the Corporate Sustainability Reporting Directive (CSRD), the companies that fall under the scope of the CSRD have to report within their yearly reports to what extent their activities are covered by the EU Taxonomy (Taxonomy-eligibility) and comply with the criteria established by the EU Taxonomy (Taxonomy-alignment).

Moreover, the International Organization for Standardization (ISO) has developed the ISO 59000 series of standards (Figs. 2.5 and 2.6) to harmonize the understanding of the circular economy and to support its implementation and measurement. Indeed, the ISO has recently published the draft of the standard: "ISO/DIS 59020 Circular economy—Measuring and assessing circularity" that aims to provide a guidance to organizations for measuring and assessing a selected system to determine its circularity performance through the adoption of proper sets of circularity indicators, considering social, environmental and economic impacts of the circular processes and actions (ISO/DIS 59020:2023).

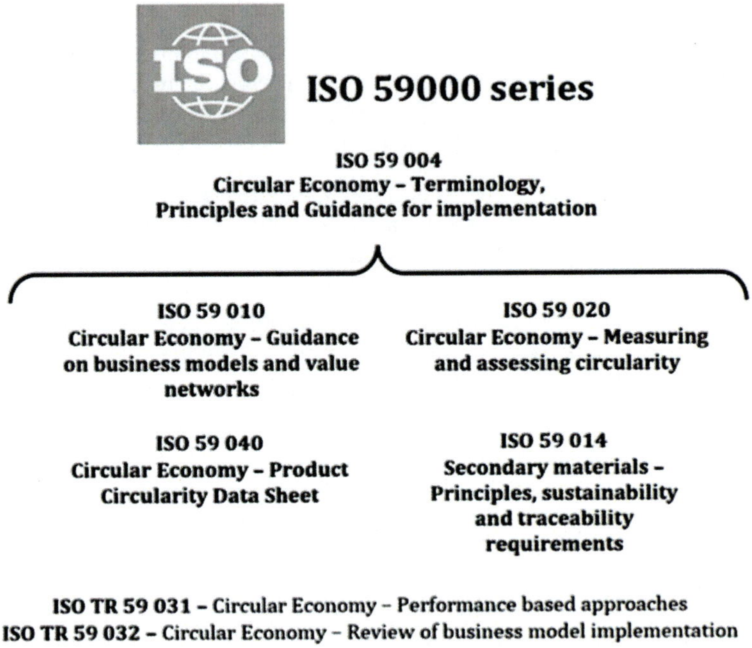

Fig. 2.5 Summary of the ISO 59000 series of standards. *Source* Sustainn (2022)

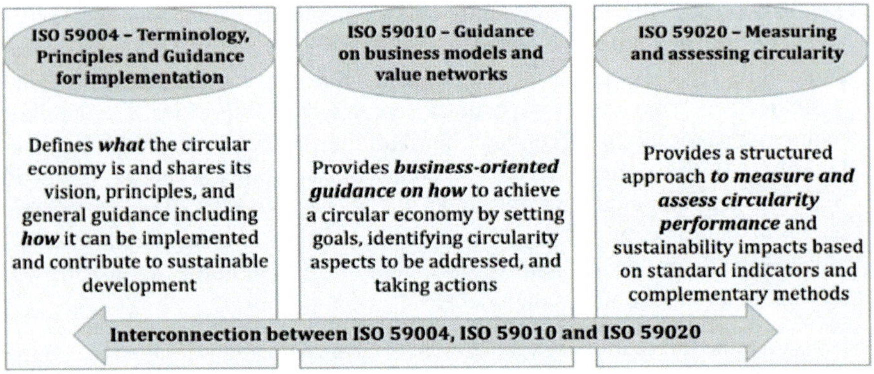

Fig. 2.6 Relationship between ISO 59004, ISO 59010 and ISO 59020. *Source* ISO/DIS 59020:2023

These support tools developed on an international and European scale demonstrate the growing attention to circularity issues in the measurement processes of strategies and actions (e.g. circular indicators, circularity assessment methods) and in assessing the level of maturity of the adoption of the principles of circularity and circular economy by companies. Therefore, the timeframe of the drafting of the sustainability reports becomes a window of opportunity to evaluate circular sustainable strategies and circularity performance in order to improve the "circular behaviour" of companies.

This topic will be deepened in Chap. 5 of the present book.

2.3 Role of Stakeholder Networks in Seizing the Opportunities of Circularity Integration Within Building Practices

As underlined in the previous paragraphs, in order to improve traditional building practices integrating circularity and environmental sustainability objectives, the contribution of stakeholders is crucial.

The cooperation among the stakeholders of the building (service and goods) supply chains represents a precondition for the integration of Circular Economy (CE) principles. Indeed, since it has a "collective nature" (Eikelenboom and de Jong 2022) CE integration requires stakeholder collaboration (Khan et al. 2020; Köhler et al. 2022). Overcoming linear approaches by adopting circular closed-loop processes implies stakeholder partnership and cross-sectoral collaboration (Hazen et al. 2021; Köhler et al. 2022) based on synergies that entail knowledge sharing and know-how pooling, according to an Open Innovation (OI)[5] approach (Köhler et al. 2022; Jesus and Jugend 2023) to realize technological innovations and develop competitive advantages (Fig. 2.7).

The interaction, collaboration and cooperation among different involved stakeholders of building processes (e.g. service providers, product suppliers, manufacturers, construction companies, facility managers, clients, building owners, property managers, designers, architects and engineers, universities and research institutes, communities, public administrations and institutions, etc.) in an Open Innovation (OI) logic can support the CE integration by contributing to an intensification of the knowledge flows (Jesus and Jugend 2023). In particular, researches by Jesus and Jugend (2023) demonstrated that «*the acquirement of knowledge and technologies*

[5] According to Chesbrough (2006), The Open Innovation paradigm represents «*the antithesis of the traditional vertical integration model where internal research and development (R&D) activities lead to internally developed products that are then distributed by the firm. […] Open Innovation is the use of purposive inflows and outflows of knowledge to accelerate internal innovation, and expand the markets for external use of innovation, respectively*» (Chesbrough 2006). Hence, the concept of Open Innovation implies that «*companies can and should use internal and external ideas to improve and accelerate their innovation processes, at the same time making their ideas, knowledge and technologies available to the external market environment*» (Jesus and Jugend 2023).

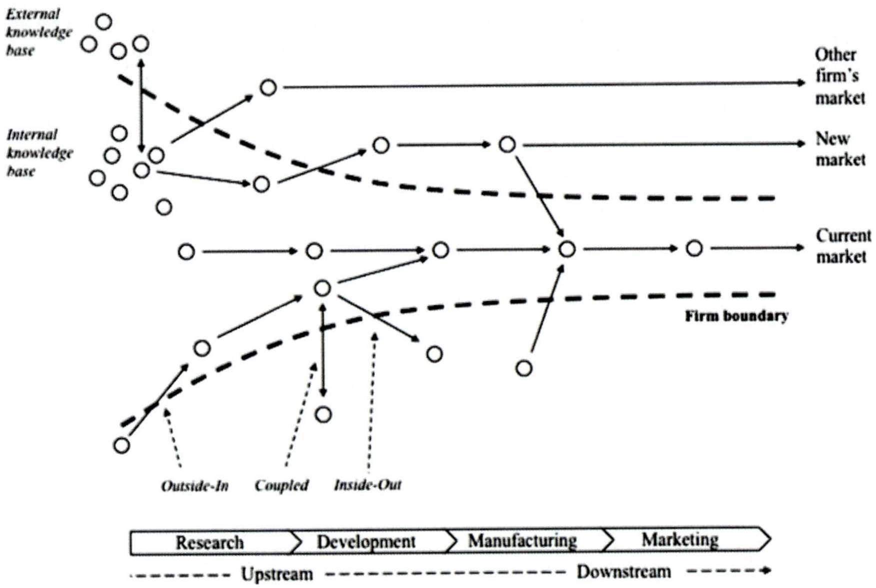

Fig. 2.7 The Open Innovation model. *Source* Chesbrough and Bogers (2014)

arising from the co-creation approach, in conjunction with stakeholder collaboration, can reduce barriers and facilitate the transition to the CE» with consequent beneficial implications (Fig. 2.8), such as: enhanced consumer awareness of the CE; co-creation of products and services leading to an increased client satisfaction; use of sharing platforms for knowledge pooling and network strengthening; company organizational performance improvement; development of smart communities; etc.

Moreover, according to Köhler et al. (2022) Open Innovation (OI) supports the creation of "dynamic capabilities" (Fig. 2.9), with mutual learning among partners.

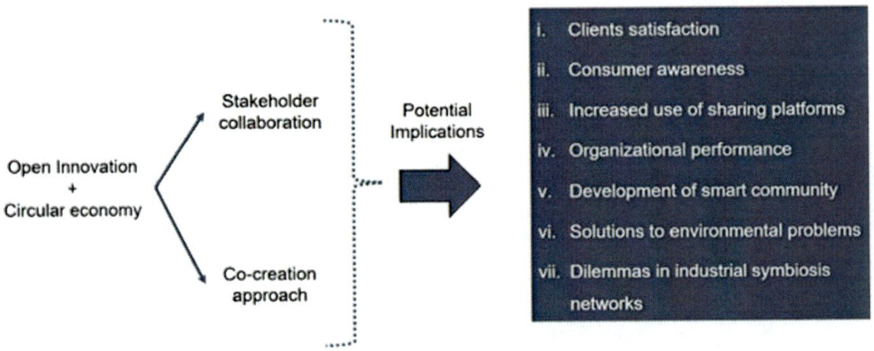

Fig. 2.8 Potential beneficial implications of Open Innovation approaches towards Circular Economy. *Source* Jesus and Jugend (2023)

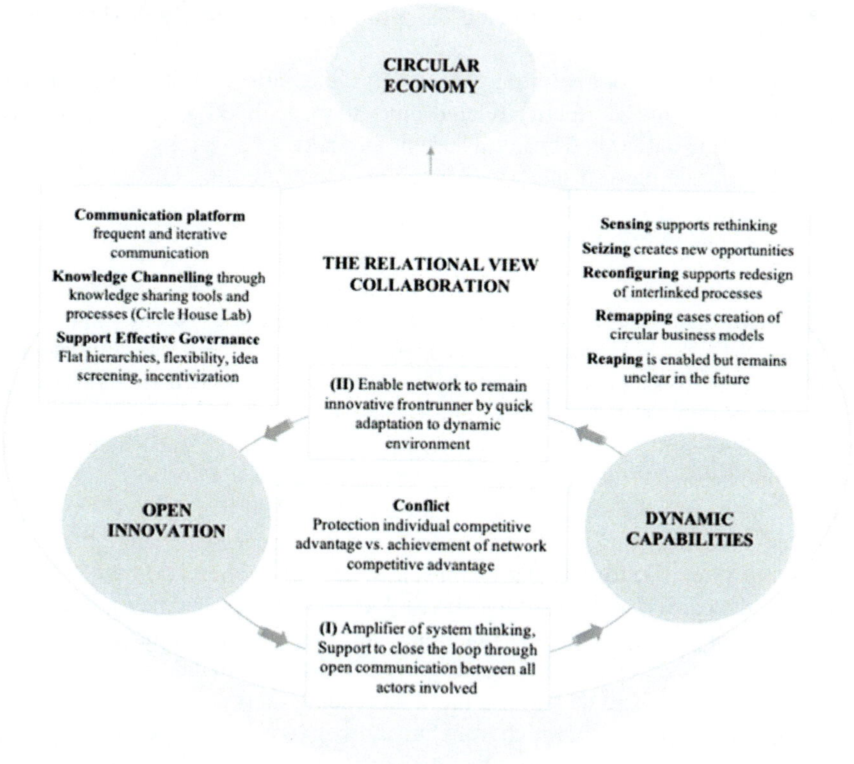

Fig. 2.9 Collaboration framework for Circular Economy: Open Innovation and dynamic capabilities. *Source* Köhler et al. (2022)

In particular:

- through inter-company communication and knowledge exchange, Open Innovation eases the access to best practices and relevant information necessary for firms to build competitive advantage;
- Open Innovation enables actors to establish business relationships with partners that own complementary know-hows «*to seize knowledge and resources by creating knowledge-sharing routines*» (Köhler et al. 2022), creating opportunities for innovations towards the gaining of new market shares. Therefore, companies are incentivized to establish in- and out-flows of information, knowledge, experiences, skills and know-hows with different external companies, creating a multidisciplinary stakeholder network because in this way they can achieve an "inter-company competitive advantage" and exploit it to boost their individual competitiveness (Köhler et al. 2022);
- the adoption of inter- and cross-company collaborative approaches leads to the creation of "supply networks", more complex and gainful than the traditional linear supply chains since they have the critical mass and resources (shared

resources) necessary to introduce CE principles into current building sector practices;

– the virtuous collaborative strategies involve the joint definition of collective CE objectives and circularity-related innovation goals by all the single stakeholders of the network. Thus, the stakeholder network resulting from the adoption of an Open Innovation approach becomes a "higher-level stakeholder" able to support the different stakeholders of the networks in gaining their individual business competitive advantages.

By a close interaction, the involved companies can define shared perspectives and collaborative strategies, capitalize on centralized knowledge, expertise and resources and share risks and responsibilities, thus being able to breaking down together the complexity of CE issues (Khan et al. 2020; Hazen et al. 2021; Eikelenboom and de Jong 2022; Köhler et al. 2022; Jesus and Jugend 2023).

In this perspective, circular practices in building processes can be seen as "incremental innovations" (Hacklin et al. 2004), that need a synergistic effort by structured multidisciplinary stakeholder networks towards a gradual review and step by step improvements of existing practices, products, service, processes, methods and tools.

The transition towards circularity, thus, involves value co-creation models developed with the participation of all the stakeholders (traditional and new) of reuse and remanufacturing processes. New functions and roles (e.g. remanufacturer, dealer, transporter/collector of post-use products, environmental manager who evaluates the construction site and decides on the waste routes, etc.) need to be integrated in the traditional supply chain, which from linear becomes a synergistic supply network of complementary companies capable of guaranteeing the circularity of processes.

References

Béland D, Howlett M (2016) The role and impact of the multiple-streams approach in comparative policy analysis. J Comp Policy Anal Res Pract 18(3):221–227

Borsboom-van Beurden J (2018) Windows of opportunity for smart city solutions in the urban fabric. In: 54th ISOCARP Congress, pp 1–5

Bos R (2020) Windows for circularity: an analysis to identify circular interventions in the different stages of the design process of an office building

Cairney P, Jones MD (2016) Kingdon's multiple streams approach: what is the empirical impact of this universal theory? Policy Stud J 44(1):37–58

Chesbrough H (2006) Open innovation: a new paradigm for understanding industrial innovation. In: Open innovation: researching a new paradigm 400:0–19

Chesbrough H, Bogers M (2014) Explicating open innovation: clarifying an emerging paradigm for understanding innovation. In: New frontiers in open innovation. Oxford University Press, Oxford, Forthcoming, pp 3–28

Cooper-Searle S (2018) Industry and policy implementation of material efficiency. Doctoral dissertation, University of Cambridge

de Graaf D, Schuitemaker S, Keita H, Vincent G (2022) Circular buildings: constructing a sustainable future. https://circulareconomy.europa.eu/platform/sites/default/files/nl-branding-circular-buildings.-f.pdf. Accessed March 2023

Derwort P, Jager N, Newig J (2022) How to explain major policy change towards sustainability? Bringing together the multiple streams framework and the multilevel perspective on socio-technical transitions to explore the German "Energiewende." Policy Stud J 50(3):671–699

Doeser F, Eidenfalk J (2013) The importance of windows of opportunity for foreign policy change. Int Area Stud Rev 16(4):390–406

EFRAG (2022) Draft European sustainability reporting standard ESRS E5 resource use and circular economy. https://www.efrag.org/lab6#subtitle4. Accessed March 2023

Eikelenboom M, de Jong G (2022) The impact of managers and network interactions on the integration of circularity in business strategy. Organ Environ 35(3):365–393

Ernst & Young - EY (2022) Corporate sustainability reporting directive. https://assets.ey.com/content/dam/ey-sites/ey-com/en_gl/topics/assurance/assurance-pdfs/ey-corporate-sustainability-reporting-directive-brochure-june-2022.pdf?download. Accessed March 2023

EU (2020) Circular economy principles for buildings design. https://ec.europa.eu/docsroom/documents/39984. Accessed March 2023

European Parliament (2022) Sustainable economy: Parliament adopts new reporting rules for multinationals. https://www.europarl.europa.eu/news/en/press-room/20221107IPR49611/sustainable-economy-parliament-adopts-newreporting-rules-for-multinationals. Accessed May 2023

EU (2023a) Public procurement for a circular economy. Good practice and guidance https://circabc.europa.eu/ui/group/44278090-3fae-4515-bcc2-44fd57c1d0d1/library/e745f67b-8c28-4174-a26a-2099b7b17789/details. Accessed May 2023

EU (2023b) Corporate sustainability reporting. https://finance.ec.europa.eu/capital-markets-union-and-financial-markets/company-reporting-and-auditing/company-reporting/corporate-sustainability-reporting_en. Accessed May 2023

Fuchs D (2017) Windows of opportunity for whom? Commissioners, access, and the balance of interest in European environmental governance. Soc Sci 6(3):73

Garrido E, Giachetti C, Maicas JP (2023) Navigating windows of opportunity: the role of international experience. Strat Manag J

Geels FW (2011) The multi-level perspective on sustainability transitions: responses to seven criticisms. Environ Innov Soc Trans 1(1):24–40

Hacklin F, Raurich V, Marxt C (2004) How incremental innovation becomes disruptive: the case of technology convergence. IEEE Int Eng Manag Con-Ference 1:32–36

Hazen BT, Russo I, Confente I, Pellathy D (2021) Supply chain management for circular economy: conceptual framework and research agenda. Int J Logistics Manag 32(2):510–537

Hernandez AG, Cooper-Searle S, Skelton AC, Cullen JM (2018) Leveraging material efficiency as an energy and climate instrument for heavy industries in the EU. Energy Policy 120:533–549

Hoefer R (2022) The multiple streams framework: understanding and applying the problems, policies, and politics approach. J Policy Pract Res 3(1):1–5

Jarvis JD, Murphy A, Perel P, Persaud N (2019) Acceptability and feasibility of a national essential medicines list in Canada: a qualitative study of perceptions of decision-makers and policy stakeholders. CMAJ 191(40):E1093–E1099

Jesus GMK, Jugend D (2023) How can open innovation contribute to circular economy adoption? Insights from a literature review. Eur J Innov Manag 26(1):65–98

Khan O, Daddi T, Iraldo F (2020) Microfoundations of dynamic capabilities: insights from circular economy business cases. Bus Strateg Environ 29(3):1479–1493

Kingdon JW (1993) How do issues get on public policy agendas. Sociol Public Agenda 8(1):40–53

Kingdon JW (2001) A model of agenda-setting, with applications. L. Rev. MSU-DCL, 331

Kingdon JW, Stano E (1984) Agendas, alternatives, and public policies. Little, Brown, Boston 45:165–169

Kingdon JW, Thurber JA (2010) Agendas, alternatives and public policies, updated Edition

Köhler J, Sönnichsen SD, Beske-Jansen P (2022) Towards a collaboration framework for circular economy: the role of dynamic capabilities and open innovation. Bus Strateg Environ 31(6):2700–2713

Lee K, Malerba F (2017) Catch-up cycles and changes in industrial leadership: windows of opportunity and responses of firms and countries in the evolution of sectoral systems. Res Policy 46(2):338–351

Lusk G (2016) The ecological footprint: new developments in policy and practice, by Andrea Collins and Andrew Flynn

Mu R (2018) Coupling of problems, political attention, policies and institutional conditions: explaining the performance of environmental targets in the national five-year plans in China. Sustain 10(5):1477

Perez C, Soete L (1988) Catching-up in technology: entry barriers and windows of opportunity. In: Dosi G, Freeman C, Nelson R, Silverberg G, Soete L (eds) Technical change and economic theory. Pinter Publishers, London, pp 458–479

Potting J, Hanemaaijer A (eds), Delahaye R, Ganzevles J, Hoekstra R, Lijzen J (2018) What we want to know and can measure. System and baseline assessment for monitoring the progress of the circular economy in the Netherlands. PBL Netherlands Environmental Assessment Agency, The Hague. https://circulareconomy.europa.eu/platform/sites/default/files/pbl-2018-circular-economy-what-we-want-to-know-and-can-measure-3216.pdf. Accessed March 2023

Sustainn (2022) ISO standards for the implementation of circular economy. https://www.wearesustainn.com/en/iso-standards-for-the-implementation-of-circular-economy/. Accessed March 2023

Tyre MJ, Orlikowski WJ (1994) Windows of opportunity: temporal patterns of technological adaptation in organizations. Organ Sci 5(1):98–118

Van Stigt R, Driessen PP, Spit TJ (2013) A window on urban sustainability: integration of environmental interests in urban planning through "decision windows." Environ Impact Assess Rev 42:18–24

Von Malmborg F, Rohdin P, Wihlborg E (2023) Climate declarations for buildings as a new policy instrument in Sweden: a multiple streams perspective. Build Res Inf 1–18

Win MD (2021) Real estate owners and the inner city. Will the shock event of the Covid-19 pandemic influence structural changes to the inner city?

Standards and Laws

BS EN 15221-1:2006 Facility management. Terms and definitions

COM (2008) 400. Communication from the Commission to the European Parliament, the Council, the European Economic and Social Committee and the Committee of the Regions. Public procurement for a better environment. https://eur-lex.europa.eu/LexUriServ/LexUriServ.do?uri=COM:2008:0400:FIN:EN:PDF

Directive 2008/98/EC. Directive 2008/98/EC of the European Parliament and of the Council of 19 November 2008 on waste and repealing certain Directives. https://eur-lex.europa.eu/legal-content/EN/TXT/?uri=celex:32008L0098

ISO/DIS 59020: 2023 Circular economy. Measuring and assessing circularity

Regulation (EU) 2020/852 of the European Parliament and of the Council of 18 June 2020 on the establishment of a framework to facilitate sustainable investment and amending regulation (EU) 2019/2088. https://eur-lex.europa.eu/legal-content/EN/TXT/PDF/?uri=CELEX:32020R0852&from=EN

Chapter 3
Requesting Circular Design Approaches: Integration of Briefing Documents (BDs) for Building Design

3.1 Level(s): Reviewing Design Approaches and Introducing New Requirements Towards Design-for-Circularity

Level(s) is the EU framework for improving the sustainability of buildings in line with the objectives of the global 2030 Agenda for Sustainable Development by United Nations (UN). Level(s) is a voluntary framework that addresses all the stakeholders involved in the building value chain. It supports building stakeholders to set targets towards circularity, by providing criteria and methods to assess and monitor the sustainability performance of buildings projects. Indeed, the framework represents an “entry point” for applying circular economy principles within building design and built environment management.

By providing a common language for assessing the circular approaches to building design, this framework represents a key data source for integrating the contents of traditional Briefing Documents (BDs) for building design with circularity-related aspects aligned with EU circularity goals. Indeed, the Level(s) framework is aligned and it interacts with key global and EU sustainability initiatives (Table 3.1).

The Level(s) framework is based on six macro-objectives which address key sustainability and circularity aspects throughout the whole building life cycle. These macro-objectives can be tracked by means of sixteen indicators (Table 3.2). In particular, the different indicators within each macro-objective describe how the building performance can be aligned with the strategic EU policy objectives in areas such as energy, water, material use, waste generation, resilience to climate change (EU 2023b).

These indicators are then characterized according to three levels (Table 3.3), following the key phases of the Building Process, namely: design, construction, use and management. Indeed, Level(s) can be applied from the very earliest phase

N. Atta, *Green Approaches in Building Design and Management Practices*, Digital Innovations in Architecture, Engineering and Construction,
https://doi.org/10.1007/978-3-031-46760-8_3

Table 3.1 Examples of how the Level(s) framework interacts with global and EU sustainability initiatives. *Source* EU (2023a)

Circularity and sustainability initiative	Interaction and alignment with Level(s) framework
Sustainable Development Goals (SDGs), Global Sustainability Development Agenda 2030	Level(s) contributes to multiple sustainable development goals (e.g. SDG13 "Climate action" and SDG11 "Sustainable cities and communities"). With particular reference to circularity in building design and management, it contributes to the SDG 12 "Ensure sustainable consumption and production patterns"
European Green Deal	Level(s) provides the bridge between the goals of the European Green Deal initiative on sustainable buildings and the realities of professional building operations within the EU
Circular Economy Action Plan	Level(s) helps understanding the full life cycle of a building and brings the circular economy into building design and use
Renovation Wave	Level(s) guides the user to apply circular economy principles to both new buildings and renovation projects. It encourages member States to base their initiatives on lifecycle thinking
Energy Performance of Buildings Directive	The Level(s) framework for sustainability performance assessments has inspired and supported the proposal for the revision of this Directive to include the assessment of whole life carbon for new buildings
Energy Efficiency Directive	The Level(s) framework for sustainability performance assessments has inspired and supported the proposal for the revision of this Directive to include the assessment of whole life carbon for the procurement of public buildings
Green Public Procurement (GPP)	Level(s) will be the basis for the revision of the different GPP criteria for office buildings. These criteria will be expanded to cover schools and social housing, and will pay particular attention to renovation
Sustainable Finance (EU Taxonomy)	Level(s) guides part of the technical screening criteria used to identify new buildings for sustainable finance
Reporting into sustainable frameworks and certifications	International sustainability certification tools are aligning their schemes to Level(s), ensuring common EU policy objectives are integrated. Level(s) enables those using this framework to report under sustainable frameworks such as the Task force on Climate-related Financial Disclosures (TCFD) and Global Real Estate Sustainability Benchmark (GRESB)
New European Bauhaus	By adopting Level(s) to assess and monitor the sustainability performance of buildings, practitioners can contribute to the New European Bauhaus agenda in the areas of the EU climate goals, circularity and healthy and comfortable spaces

Table 3.2 Level(s): sets of indicators for the six macro-objectives. *Source* EU (2023b)

Level(s) macro-objective		Macro-objective description	Indicator	
1	Greenhouse gas emissions along a building life cycle	Minimize the whole life carbon output, consider both energy consumption during the use phase of the building and embodied energy	1.1	Use stage energy performance (kWh/m^2/yr)
			1.2	Life cycle Global Warming Potential (CO_2 eq./m^2/yr)
2	Resource efficient and circular material life cycles	Optimize the building design to support lean and circular flows, including: – Building materials use and quantities – Minimize construction and demolition waste generated to optimize material use – Replacement cycles and flexibility to adapt to change – Potential for deconstruction as opposed to demolition	2.1	Bill of quantities, materials and lifespans
			2.2	Construction and demolition waste and materials
			2.3	Design for adaptability and renovation
			2.4	Design for deconstruction, reuse and recycling
3	Efficient use of water resources	Use water efficiently, particularly in areas of identified long-term or projected water stress	3.1	Use stage water consumption (m^3/occupant/yr)
4	Healthy and comfortable spaces	Create buildings that are comfortable, attractive and productive. This includes four aspects of the quality of the indoor environmental quality: – The indoor air for specific parameters and pollutants – The degree of thermal comfort – The quality of artificial and natural light and associated visual comfort – The capacity of the building fabric to insulate occupiers from internal and external sources of noise	4.1	Indoor air quality
			4.2	Time outside of thermal comfort range
			4.3	Lighting and visual comfort
			4.4	Acoustics and protection against noise

(continued)

Table 3.2 (continued)

Level(s) macro-objective		Macro-objective description	Indicator	
5	Adaption and resilience to climate change	Futureproof building performance: – Adapt to changes of future climate impacting on thermal comfort – Make the building more resilient and resistant to extreme weather events (including flooding: fluvial, pluvial and coastal) – Improve the building design to reduce the chances of pluvial/fluvial flood events in the local area (i.e. increasing sustainable drainage)	5.1	Protection of occupier health and thermal comfort
			5.2	Increased risk of extreme weather
			5.3	Sustainable drainage
6	Optimized life cycle cost and value	Long term view of the whole life costs and market value of more sustainable buildings, including: – Life cycle costs (construction, operation, maintenance, refurbishment and disposal) – Encourage the integration of sustainability aspects into market value assessment and risk rating processes and ensure that this is done as informed and transparent as possible	6.1	Life cycle costs (€/m^2/yr)
			6.2	Value creation and risk factors

of conceptual design (setting objectives at concept stage), till the in-use phase of the building.

In particular, according to EU (2023c):

- *Level 1. The conceptual design for the building project.* This first level is the simplest of the three and it involves early-stage qualitative assessments of the sustainability and circularity propensity of the conceived concept (conceptual design);
- *Level 2. The detailed design and construction performance of the building.* This second level is the intermediate one and it entails the quantitative assessment of the integration of sustainability and circularity principles within the detailed design project (designed performance) as well as the monitoring of the planned construction activities according to standardized criteria and methods.
- *Level 3. The as-built and in-use performance of how the building performs after completion and handover to the Client.* This third and last level is the most complex one and it involves firstly the surveying of the as-built and secondly the monitoring overtime of the in-use behavior of the building and its components to assess circular strategies and accordingly adjust them if needed according to a continuous improvement approach.

Hence, scaling the macro-objectives and related indicators introduced in Table 3.2 according to the three levels of advancement of the building process stages, it is possible to specifically apply the indicators to each key stage (i.e. design, construction, use and management), as shown in Table 3.4.

Table 3.3 The three levels of the EU Level(s) framework. *Source* EU (2023c)

Level no	Level title	BP phase and level objective	Level description (EU 2023c)
Level 1	Conceptual design	Setting objectives at concept stage	Early-stage qualitative assessments and reporting on the concepts that the chosen indicators will cover. It provides a simple structure that can be presented to clients to prioritize attention on sustainability aspects
Level 2	Detailed design and construction	Assess performance at design and construction	Quantitative assessment of the designed performance. Allowing comparison between different design options and monitoring of the construction according to standardized units and methods
Level 3	As-built and in-use	Follow up after completion	Monitoring and surveying of activity both on the construction site and of the completed building and its first occupants. Level 3 supports building operators in understanding the actual building performance and in identifying the lessons learned from the design to inform and improve future projects

Among all the Level(s) indicators (Tables 3.2 and 3.4), the indicators 2.2, 2.3, 2.4 are the ones that mostly concern the issues of circularity and closed cycles of use and management of resources. These indicators and their articulation in the Level 1 (conceptual design) are particularly useful for integrating the contents of current Briefing Documents (BDs) for building design. In particular the EU framework also proposes User Manuals and Reporting Templates to support the integration of circularity specifications and requirement in the project design. In this regard, the following sub-paragraphs highlight the contribution of these tools for applying the above-mentioned key indicators for circularity to the design stage (Conceptual Design—Level 1; Detailed design—Level 2) of the Building Process.

3.1.1 EU Level(s) Indicator 2.2 "Construction and Demolition Waste and Materials"

According to the User Manual of EU Level(s) Indicator 2.2 "Construction & Demolition waste and materials" (EU 2021a), at the preliminary design stage (Level 1) the main stakeholders to involve in the process of integration and assessment of circular approaches are: the building authority, the architect, the building owner and other investors of the construction/renovation/demolition work together with the contractor, waste managers and product manufacturers (Fig. 3.1).

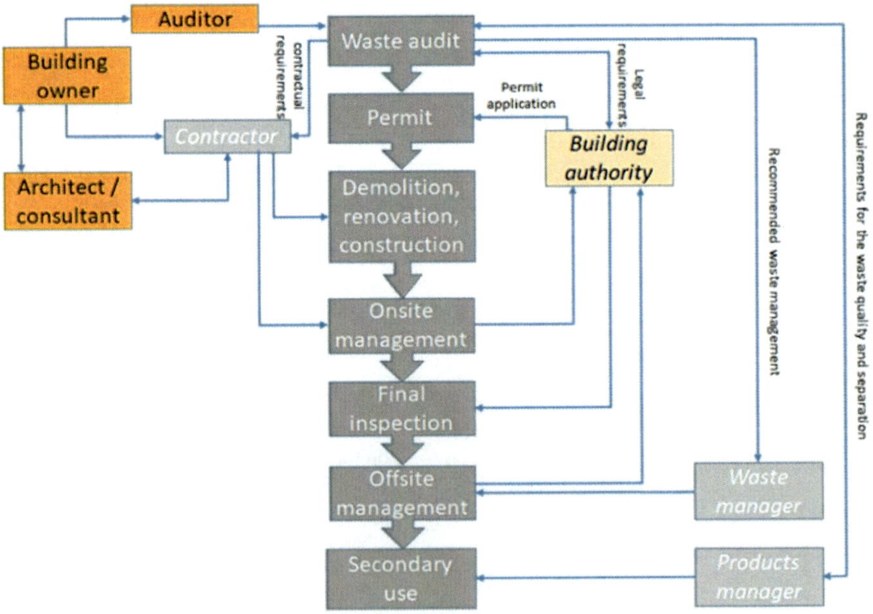

Fig. 3.1 Building project stakeholders: roles and relationships. *Source* EU (2021a)

Table 3.4 EU Level(s) indicators applied to the three key stages of the Building Process. *Source* EU (2023c)

Macro-objective 1. Greenhouse gas emissions along a buildings life cycle	
Indicator 1.1 Use stage energy performance (kWh/m²/yr)	
Level 1 Conceptual design	*Aims*: (i) understand the energy needs associated with the type of building they are working on, and (ii) know where to focus the attention to reduce the total primary energy use associated with the building's delivered energy needs during the use stage *Tool*: checklist to inform on design energy concepts
Level 2 Detailed design and construction	*Aim*: calculate the energy needs and primary energy use of a building for the purpose of design comparisons, building permitting or tendering *Tool*: energy simulation softare
Level 3 As-built and in-use	*Aims*: (i) collect metered data to understand the energy needs associated with the building; (ii) carry out testing of the building in use to identify any performance issues with the building fabric and technical services *Tool*: energy meter for real time measuring of energy consumption
Indicator 1.2 Life cycle global warming potential (CO_2 eq./m²/yr)	
Level 1 Conceptual design	*Aims*: (i) incorporate some important life cycle concepts into design and, later, into detailed designs; (ii) interpret and use the results of previously carried out life cycle GWP assessments and life cycle assessments that are based on the analysis of similar building types *Tool*: checklist to integrate life cycle concepts
Level 2 Detailed design and construction	*Aims*: (i) calculate the life cycle GWP emissions of their project and select software tools and databases according to the standard EN 15978 *Tool*: 'hot spot' analysis and related results interpretation and use
Level 3 As-built and in-use	The same procedure as defined in Level 2 but supported by the certainty of materials procured and technical building systems installed
Macro-objectve 2. Resource efficient and circular material life cycles	
Indicator 2.1 Bill of quantities, materials and lifespans	
Level 1 Conceptual design	Focus on the six highly relevant aspects for optimising the consumption of construction materials and products; (ii) describe how these aspects were considered (or not) during discussions and decision-making at the concept design stage
Level 2 Detailed design and construction	– Make an estimate of bill of quantities (BoQ) during the design stage that ensures that budgetary limits are respected – Use an inventory template to insert and manage the BoQ data. Furthermore, by entering optional cost data and lifespans, the BoQ template can generate outputs that are useful for other Level(s) indicators
Level 3 As-built and in-use	– Register and log BoQ data as materials and products are procured and delivered to the site based on actual quotations and purchases – Use an inventory template to centralize record of purchases to track spending in line with project budgets and schedules – Compare with estimates during design stage
Indicator 2.2 Construction and Demolition Waste (CDW) and materials	

(continued)

Table 3.4 (continued)

Level 1 Conceptual design	– Be aware of highly relevant aspects for reducing CDW and optimising its management – Describe how these aspects were considered (or not) during discussions and decision-making at the concept design stage
Level 2 Detailed design and construction	– Report on and to make reliable quantitative estimates of CDW – Use (an) inventory template(s) for CW and/or DW estimation
Level 3 As-built and in-use	– Measure the quantities of CDW in their project, using the Level(s) excel templates for CW and DW reporting to collect data – Compare estimates with actual data
Indicator 2.3 Design for adaptability and renovation	
Level 1 Conceptual design	– Understand how the design of a building could facilitate future adaptation to changing occupier needs and market conditions – How these design aspects could extend the service life of the building as a whole, either by facilitating continuation of the intended use or through possible future changes in use
Level 2 Detailed design and construction	Set design targets and compare design options for their relative adaptability
Level 3 As-built and in-use	Compare the final as-built design with the earlier detailed designs. It can also form the starting point for a long-term monitoring of the building and how it performs in the local property market
Indicator 2.4 Design for deconstruction, reuse and recycling	
Level 1 Conceptual design	– Understand how the design of a building could facilitate ease of future deconstruction in order to access, disassemble and dismantle parts and materials – Consider the extent to which the building parts may be recovered for either reuse and/or for recycling
Level 2 Detailed design and construction	Set design targets and compare design options for their deconstruction potential
Level 3 As-built and in-use	Compare the final as-built design with the earlier detailed designs. It can also form the starting point for preparing the technical content of a building passport or building material bank record
Macro-objective 3. Efficient use of water resources	
Indicator 3.1 Use stage water consumption (m^3/occupant/yr)	
Level 1 Conceptual design	– Be aware of five highly relevant aspects for reducing and optimising use stage water consumption – Describe how these aspects were considered (or not) during discussions and decision-making at the concept design stage – Prepare a checklist to inform on water efficient design concepts

(continued)

Table 3.4 (continued)

Level 2 Detailed design and construction	– Estimate the per person water consumption in the building as a function of the water consuming devices, appliances and irrigated areas via an excel-based calculator – Minimize potable water consumption by the specification of more efficient devices and appliances and by rainwater harvesting and/or greywater reuse
Level 3 As-built and in-use	– Take measures of actual water consumption over the course of one year – Estimate occupation rates of the building – Compare estimates with measures
Macro-objective 4. Healthy and comfortable spaces	
Indicator 4.1 Indoor air quality (IAQ)	
Level 1 Conceptual design	– Be aware of three highly relevant design aspects that represent the main factors that influence IAQ and contribute to optimising the ventilation strategy for a building – Describe how these aspects were considered (or not) during discussions and decision-making at the concept design stage
Level 2 Detailed design and construction	– Design the ventilation system, the specification of indoor fit-out materials and, in the case of major renovations, the design of insulation and other design improvements to the air tightness and integrity of the building fabric – Quantify the ventilation rates needed in different zones of the building – Factor into the design the potential influences on the quality of outdoor air (e.g. proximity of roads, traffic volume, etc.) and on the quality of indoor air (e.g. emissions from materials, bio-effluents, point sources of humidity, etc.) – Inform ventilation system specifications. The difference in quality between outdoor air entering the system and the desired quality of the air to be supplied indoors will directly influence the filter specification, which in turn will influence the sizing of the system and its energy performance
Level 3 As-built and in-use	– Assess IAQ in an objective manner based on the performance of a completed building, two-pronged approach recommended: a. Quantitative approach based on air sampling and monitoring at two different stages: (i) after completion but prior to occupation and (ii) during occupation b. Qualitative approach based on occupant feedback in surveys about IAQ during occupation and a remedial action plan in case of unsatisfactory results
Indicator 4.2 Time outside of thermal comfort range	
Level 1 Conceptual design	– Assess the risks of occupier thermal discomfort during the heating and cooling seasons for the building type being assessed – Understand measures that can be taken to create a comfortable thermal environment in the building types being assessed

(continued)

Table 3.4 (continued)

Level 2 Detailed design and construction	– Assess the energy requirements of a building – Make a quantitative assessment of the indoor thermal conditions according to the Category II temperature ranges stipulated in EN 16978-1 (or national equivalent) – Make an overheating assessment of a building for the purpose of obtaining a building permit
Level 3 As-built and in-use	– Collect monitoring data on the thermal conditions in a building to compare the performance with design simulations, and/or – Carry out a post-occupancy survey of occupants to determine the level of dissatisfaction with the thermal comfort conditions and compare the results with the design estimates
Indicator 4.3 Lighting and visual comfort	
Level 1 Conceptual design (Single Level)	– Support in understanding and prioritising the most important aspects of lighting and visual comfort to focus attention on – Setting requirements and specifications that facilitate the later detailed design that is supportive of occupant health and comfort during visual tasks and activities
Indicator 4.4 Acoustics and protection against noise	
Level 2 Conceptual design (Single Level)	– Understand and prioritize the most important design aspects to focus attention on considering five acoustic and noise protection design aspects at the concept stage of a project – Setting requirements and specifications. This level also provides initial suggestions for how calculations and field measurements can be made
Macro-objective 5. Adaption and resilience to climate change	
Indicator 5.1 Protection of occupier health and thermal comfort	
Level 1 Conceptual design	– Assess the risks of occupier thermal discomfort during the cooling seasons for the building type being assessed – Understand and identify measures that can be taken to future-proof a building thermal environment and/or incorporate adaptation measures
Level 2 Detailed design and construction (Last Level)	Assess the energy requirements of a building and make a quantitative assessment of the indoor thermal conditions under projected future climate conditions
Indicator 5.2 Increased risk of extreme weather	
Level 1 Conceptual design (Single Level)	– Be aware of steps to take during the conceptual design stage (and even earlier) to ensure that the awareness of extreme weather events at the building location is maximised – Optimize the design of the building and any surrounding plot area for adaptation to extreme weather events
Indicator 5.3 Sustainable drainage	
Level 1 Conceptual design (Single Level)	– Set out the steps to take during the conceptual design stage in order to embrace sustainable drainage options as much as possible – Be aware of both the risk of flooding at the building and the possible effect of the building itself on flood risk in surrounding and downstream areas
Macro-objective 6. Optimized life cycle cost and value	

(continued)

Table 3.4 (continued)

Indicator 6.1 Life cycle costs (€/m²/yr)	
Level 1 Conceptual design	– Calculate the life cycle cost of their building project and understand how to take a longer-term perspective on the costs associated with a building project – Incorporate some important life cycle cost concepts into conceptual designs and, later, detailed designs
Level 2 Detailed design and construction	– Calculate the life cycle costs of their building project – Select software tools and databases – Understand the basic parts of the calculation and the calculation steps according to the cost optimal method, EN 15459 and standard ISO 15686-5, including assumptions and default parameters that shall be used and data gap filling
Level 3 As-built and in-use	– Revise the life cycle costs of their building project based on the as-built initial costs and any associated revisions in the projected annual and periodic costs – Report on the life cycle costs for a completed building
Indicator 6.2 Value creation and risk factors	
Level 1 Conceptual design (Single Level)	– Understand how sustainability aspects addressed by Level(s) can have an influence on property financial value and risk appraisals – Have a starting point for dialogue between the design team, the client and their property market specialists – Ensure that improved sustainability performance, which can be reported on and verified using Level(s), is among the factors taken into account in making a property market valuation

The involved stakeholders have to jointly discuss about the integration of "circular design concepts" (EU 2021a) in the process of development of the concept building design project. In particular, according to the EU Level(s) Indicator 2.2 (EU 2021a), the key "circular design concepts" related to the topic "Construction and Demolition waste and material" to introduce are referred to/pursued through (EU 2021a):

1. Setting of relevant targets of Key Performance Indicators (KPIs);
2. Influence of project type on CWD generation and management;
3. Pre-demolition audit;
4. Good construction practice;
5. Outline Waste Management Plan (WMP);
6. "Building As Material Banks" (BAMB) principles.

In particular, Table 3.5 describes these 6 points and provides examples of reporting.

The Level 2 involves the accurate quantification and characterization of construction and/or demolition waste/scrap typologies and the definition of the actions to activate in order to prepare the waste for reuse, remanufacturing, recovery or recycling. To this end, the User Manual of the EU Level(s) Indicator 2.2 (EU 2021a) provides a standardized procedure and templates.

Table 3.5 EU Level(s) Indicator 2.2 Design concept (Level 1): features and reporting. *Source* Adapted from EU (2021a)

EU Level(s) indicator 2.2 "Construction and demolition waste and material"		
Level 1 design concept	Description	General examples of reporting type/method/contents
1. Setting of relevant targets of key performance indicators (KPIs)	Definition of goals expressed in measurable "ambition levels" (targets) and related indicators for the measurement/evaluation of the actual achievement of the predefined targets related to the waste management process. The ambition levels (and related KPIs) should be aligned with the principles of the EU waste hierarchy and the most updated regulatory framework on the subject (e.g. EU taxonomy contribution criteria and DNSH targets).	First preliminary rough estimation of the quantities and characterization of waste according to the "EU List of Wastes". Defined targets for non-hazardous waste management according to the EU waste hierarchy and aligned with the EU taxonomy: • $\leq$5% landfill disposal; • $\geq$95% material/inert recovery, recycling or reuse.
2. Influence of project type on CWD generation and management	Definition of the project type (e.g. new construction on greenfield site, extraordinary maintenance works, renovation project or demolition project) and its influence on the waste generation (e.g. different projects generate different quantities of waste).	Definition of the project type: (i) new-building construction in rural area, in sub-urban area, in urban area, on cleared brownfield site; (ii) demolition and new construction, thus the waste can vary significantly depending on the differences between the old and the new building; (iii) and/or renovation. Conduction of a preliminary analysis of the quantities and types of waste related to the type of building works. In all the cases, the different opportunities to maximise the reuse of building elements and the recycling or the recovery of materials should be outlined and assessed.

(continued)

Table 3.5 (continued)

EU Level(s) indicator 2.2 "Construction and demolition waste and material"

Level 1 design concept	Description	General examples of reporting type/method/contents
3. Pre-demolition audit	Performing of a pre-demolition audit for demolition or renovation projects in order to identify the different systems, elements, products, materials and potential scraps/waste and how they can be recovered, reused, remanufactured and/or recycled (or landfilled as last option), as well as to plan the needed activities of selective deconstruction, stripping and sorting techniques, also in relation with available time, space and labour (i.e. key constraints of the project that has to be assessed).	For projects that involve demolition activities, it is advisable/required (according to the national regulations and requirements) the development of a report of the pre-demolition audit. The report of the pre-demolition audit should include an inventory of construction elements and materials within the old building/infrastructure, including their identification, descritption and localization within the building (accessibility and ease to disassembly) with particular attention to hazardous materials. Note: The EU Level(s) already provide a common template for this inventory. In order to define the proper re-strategy (according to the EU waste hierarchy) it is important that the inventory highlights: – How the systems/elements/products/materials can be removed in a safe and environmentally sound manner with minimal compromise of technical quality for future reuse, recycling or recovery; – Definition of the possibile second uses of the identified elements including assessment for internal reuse (use of the identified elements for the new building under design) or external reuse. In this second case, it is advisable/required to identify the possible sale-market and organizational models (e.g. product-as-a-service) for each inventoried element. When the recovery is not possible, the best solution, in terms of environmental sustainability, for the disposal must be identified and assessed based on local and regional facilities/infrastructures/specificity.

(continued)

Table 3.5 (continued)

EU Level(s) indicator 2.2 "Construction and demolition waste and material"		
Level 1 design concept	Description	General examples of reporting type/method/contents
4. Good construction practice	Definition of construction approaches and practices that can reduce onsite Construction Waste generation (e.g. prefabrication) and procurement arrangements that may incentivize the reduction of over-ordered materials.	Indication of the construction approaches, strategies, methods and practicies towards circularity and waste prevention, such as for instance: – Adoption of modular construction elements with standardized dimensions; – Priority to the selection of prefabricated elements; – Incorporation of a Waste Management (WM) expert into the project team for a training on the WM topic and how to develop a WM Plan for the project; – Provision of onsite training to contractors and sub-contractors; – Planning of monitoring and progress meetings about site waste during the construction phase; – Define possible ways to incentivize the contractor to minimize over-ordered material (e.g. via specific KPIs and contractual clauses, inlcuding for instance new buy-back options with suppliers for over-ordered materials)

(continued)

Table 3.5 (continued)

EU Level(s) indicator 2.2 "Construction and demolition waste and material"		
Level 1 design concept	Description	General examples of reporting type/method/contents
5. Outline waste management plan (WMP)	Development of a preliminary WMP—Waste Management Plan (to be further updated and articulated in the subsequent stages of the building process) that describes: – How environmental and health impacts from CDW can be reduced; – How to manage hazardous wastes on site; – How to separate the collection of different types of non-hazardous CDW as a function of planned treatment options, which inlude: cleaning for reuse; reuse (e.g. structural steel, metal sheet, roof tiles); recycling (e.g. metals, glass); material recovery for repurposing (e.g. bricks to aggregates, wood to particleboard); energy recovery (e.g. wood, plastics, certain insulation materials); disposal (e.g. landfill or thermal destruction without energy recovery); – How the selective collection of onsite waste can be optimised based on the different possible end-market, storage, processing and disposal options; – How cost benefits can be maximised (i.e. increased revenues and avoided costs).	The WMP should contain information concerning: – How and by whom the different steps of the demolition will be performed; – Which materials will be collected selectively and where and how they will be transported; – Related costs, including approximate per tonne disposal costs (e.g. landfill gate fees and taxes), treatment costs (e.g. €/tonne to crush and grade concrete) and avoided material costs (e.g. €/tonne virgin aggregate for backfill); – Final outcome for each waste stream; – Measures taken to limit environmental impacts during the waste generating activities, waste storage and waste transport (e.g. including leaching and dust); – Procedures to best manage hazardous waste (if any).

(continued)

Table 3.5 (continued)

EU Level(s) indicator 2.2 "Construction and demolition waste and material"		
Level 1 design concept	Description	General examples of reporting type/method/contents
6. "Building as material banks" (BAMB) principles	Assess how the BAMB concepts could be applied to the conceptual design of the building (e.g. design for disassembly and reuse of prefabricated elements at the end-of-life both in case of the design of new buildings or in the definition of renovation activities). The possibility to reuse building elements can be maximised by choosing products and components that can be easily disassembled and by providing clear instructions about their correct dis/assembly (e.g. through Product Passports that can include product features, reuse/remanufacturing/recycling potential and visual details of the building element)	It is advisable to include within the design team an expert of the BAMB principles that can support the design team in the development of Product Passports. The passports can be digital (Digital Product Passports) and they can also be linked to the 3D representation of the building elements and the related sets of available data and documents within the BIM model (Building Information Modelling) of the building (if available/to be developed in the design stage).

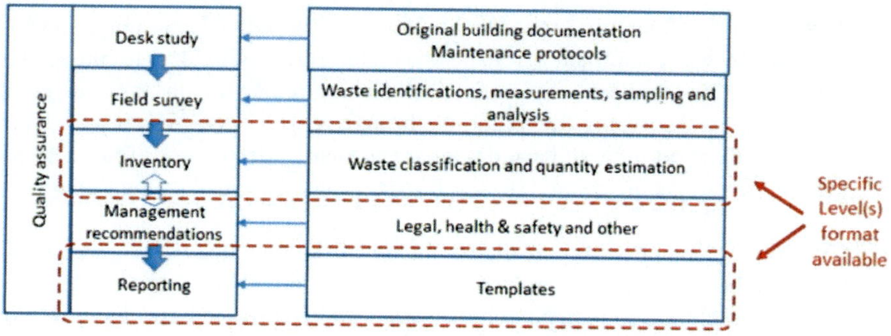

Fig. 3.2 The key steps for reporting on demolition waste by EU Level(s) Indicator 2.2. *Source* EU (2021a)

In particular, the procedure proposed in the Level 2 by the User Manual for the demolition activities is articulated into the following steps (Fig. 3.2):

1. Documental audit, namley the collection and review of the available original building documentation to make an initial estimate of the involved products and materials;
2. Field survey, in order to integrate and validate the data and information callected with the activity at point 1 (identification and quantification of products and materials). In addition, the involvement of specialist contractors with knowledge of local/regional reuse-markets is advisable in order to define the proper re-strategies for the types of identified products/materials;
3. Inventory as outcome of pre-demolition audits. Following the Level(s) template for Demolition Waste (DW) estimates and the instructions embedded in the Level(s) inventory template it is possible to develop an inventory of resource/waste estimates related to demolition/renovation activities. Hence, this phase is aimed at concretize the outcome of the activities in points 1 and 2, also concluding the pre-demolition audit with the development of the inventory and setting the basis for the subseuent outlining of the Waste Management Plan (WMP);
4. First draft of WMP, that describes how the elements, materials and waste deriving from the demolition activity should be collected, stored, treated and transported and how data on the waste and resources (product/material passports) arising on the site will be monitored and updated overtime.

For what concerns the construction activities, according to the User Manual of the EU Level(s) Indicator 2.2 (EU 2021a), the related procedural steps of Level 2 are:

1. Drafting of a first estimatation of Bill of Materials (BoM) and Bill of Quantities (BoQ) for the project, based on the detailed design documentation and related data input within the Level(s) template for Construction Waste (CW) estimates;
2. Characterization of the potential waste material (e.g. inert, non-hazardous or hazardous) and identification of the appropriate waste code;

3. Analysis aimed at identifying the most suitable and sustainable circular processes proposed by waste contractors for specific material/waste streams (for preventing the generation of waste by activating reusing/remanufacturing/repurposing processes, by recovering the materials through recycling or—if not possible—for waste disposal in landfill) and related end-market destinations (based on the local/regional specific context) for each waste material (traced in point 2) in order to complete the CW inventory;
4. First draft of WMP based on the outcomes of the previuos points. The WMP describes how the elements, materials and wastes deriving from the construction activity should be collected, stored, treated and transported and how data on the waste and resources (product/material passports) arising on the site will be monitored and updated overtime.

3.1.2 EU Level(s) Indicator 2.3 "Design for Adaptability and Renovation"

The level 1 of the Indicator 2.3 "Design for adaptability and renovation", according to the User Manual of EU Level(s) (EU 2021b), is aimed at integrating in the concept design processes the key circular design approaches to extend the useful-life of the building elements, to lengthen the service-life of building elements (single use-cycle that matches the useful-life) and/or to maximize the post-first-use usecycles of building elements (multiple consequent use-cycles with same or different element function).

In particular, the level 1 involves a set of key "adaptability design concepts" for the design of buildings "adaptable" to future changes in use with improved long-term environmental performance. The User Manual of the Indicator 2.3 provides two sets of "adaptability design concepts", one applicable to office buildings (Table 3.6) and one applicable to residential buildings (Table 3.7).

Furthermore, the level 2 supports the setting of design targets and the comparison of design options for their relative adaptability by providing methods to give scores to each "design concept" introduced in Tables 3.6 and 3.7. To this end, the User Manual provides the following suggestions (EU 2021b):

- define the different design options taking into account different combinations of adaptability design aspects for a subsequent joint assessment by architects, structural and service engineers, property market specialists (with knowledge of the local/regional market needs) and other relevant stakeholders at this concept design stage;
- *"identify and gather all the relevant architectural and structural design drawings, service plans as well as supporting calculations required to make the scoring" [...] "in ordert to obtain the adaptability score for the design, multiply the score obtained for each design aspect by the weighting factor and then sum up the weighted scores, to obtain a score out of 100"* (EU 2021b);

Table 3.6 Adaptability design concepts for office buildings by EU Level(s) Indicator 2.3. *Source* EU (2021b)

Adaptability design concept	Specific design aspect to address	How the design aspect can contribute to adaptability (EU 2021b)
1. Changes to the internal space distribution	1.1 Column grid spans	Wider column spans will allow for more flexible floor layouts
	1.2 Façade pattern	Narrower bays will allow for more internal space configurations
	1.3 Internal wall system	Non-loading bearing internal walls will allow for changes to be more easily made to floor layouts
	1.4 Unit size and access	By ensuring that access/egress is possible from sub-divisions of the spaces this will provide more sub-letting options
2. Changes to the buildings servicing	2.1 Ease of access to service ducts	Access will be improved if services are not embedded in the building structure
	2.2 Ease of access to plant rooms	Future changes of technical equipment will be facilitated if there is ease of access to plant rooms and equipment
	2.3 Longitudinal ducts for service routes	The inclusion of longitudinal ducts will provide flexibility in the location of service points
	2.4 Higher ceilings for service routes	The use of greater ceiling heights will provide more flexibility in the routing of services
	2.5 Services to sub-divisions	By ensuring that individual servicing for sanitary facilities is possible for sub-divisions of the spaces, this will provide more sub-letting options
3. Changes to the building façade and structure	3.1 Non-load bearing facades	Non-load bearing facades will allow for changes to be made more easily to both internal layouts and external elements

(continued)

Table 3.6 (continued)

Adaptability design concept	Specific design aspect to address	How the design aspect can contribute to adaptability (EU 2021b)
	3.2 Future-proofing of load bearing capacity	The incorporation of redundant load bearing capacity will support potential future changes in the building façade and uses
	3.3 Structural design to support future expansion	Structural designs that have the vertical strength to support additional storeys will allow for future expansion of the floor area

- involve energy/environment/sustainability experts to assess the life-cycle performance of the different building design options in order to identify possible trade-offs in the life cycle environmental performance of the different adaptability design concepts. To this end, a life cycle GWP (Global Warming Potential) assessment or LCA (Life Cycle Assessment) can be carried out.

3.1.3 EU Level(s) Indicator 2.4 "Design for Deconstruction, Reuse and Recycling"

The level 1 of the Indicator 2.4 "Design for deconstruction, reuse and recycling", according to the User Manual of EU Level(s) (EU 2021c), is aimed at understanding how choices in the design stage of a building can significantly increase the ability of building parts to be easily disassembled (ease to disassembly), maintained and repaired and/or replaced (maintainability) overtime and, when possible, dismantled into their materials (EU 2021c). The "deconstruction design concepts" for designing easy-to-disassembly solutions and assessing the extent to which the building elements can be recovered for either reuse and/or recycling proposed by the User Manual of the Indicator 2.4 are reported in Table 3.8.

The level 2, on the basis of the "deconstruction concpets" highlighted in Table 3.8, is aimed at setting quantitative design targets supporting the decision-making in the design stage. The purpose is to compare design options for their deconstruction potential based on quantitative scoring of design-for-deconstruction.

To this end, the level 2 proposes a set of instructions as follows (EU 2021c):

- create a multidisciplinary design team, gathering together architect/s, structural engineers, service engineers, fit-out contractors, environmental consultants, demolition contractors and waste management experts in order to review the design concepts, construction and deconstruction techniques and the local/regional end-markets for elements to reuse/recycle (from design drawings, bill of quantities and materials, fit-out plans);
- identify possible viable re-strategies for the different building elements, taking into account the different deconstruction design concepts and aspects and if these

Table 3.7 Adaptability design concepts for residential buildings by EU Level(s) Indicator 2.3. *Source* EU (2021b)

Adaptability design concept	Specific design aspect to address	How the design aspect can contribute to adaptability (EU 2021b)
1. Changes to the internal space distribution	1.1 Wall systems that support layout changes	Internal wall designs that allow for ease of changes to floor layouts
	1.2 Greater ceiling heights for surface routes	The use of greater ceiling heights will give more flexibility in the routing of services
2. Changes to the buildings servicing	2.1 Ease of access to the building services	Location of services within the house or apartment building to ensure they are flexible to change
	2.2 Ease of adaptation of the distribution networks and connectors	The internal layout of rooms, the distribution networks and connections can be adapted in the case of any layout modifications
3. Change to the use of units or floors	3.1 The potential for a segregated home working spaces	The potential to segregate a space with adequate dimensions, light and services within the home will support teleworking
	3.2 The potential for ground floor conversion to a contained unit	The potential for the ground floor to become a contained unit with bed space, kitchen, toilet and shower will support future changes in circumstances
4. Changes in access requirements	4.1 Ease of access to each residential unit	Ease of access to residential units in cases of the need for pram or wheelchair mobility
	4.2 Access to and manoeuvrability within rooms	Ease of access to and space to manoeuvre within living spaces, kitchens and bathrooms will support pram or wheelchair mobility

Table 3.8 Deconstruction design concepts for residential buildings by EU Level(s) Indicator 2.4. *Source* EU (2021c)

General deconstruction aspect	Specific design aspect to address	Description (EU 2021c)
1. Ease of disassembly	1.1 Elements and their parts are independent and easily separable	Potential to separate building elements that are connected to each other and to disassemble elements into their constituent components and parts
	1.2 Connections are mechanical and reversible	Use of mechanical, non-destructive connections as opposed to chemical bonding
	1.3 Connections are easily accessible and sequentially reversible	Easy and sequential access in order to reverse mechanical connections and remove elements, components or parts
	1.4 The number and complexity of the disassembly steps are low	The disassembly should not suppose the need for complex preparatory steps, the intensive use of manpower and machinery and/or off-site processes
2. Ease of reuse	2.1 Specification of elements and parts using standardised dimensions	Specification of elements and parts that are of a standardised specification in order to provide consistent future stock
	2.2 Specification of modular building services	Specification of modular systems that may retain value upon de-installation or which may be more easily swapped out and upgraded
	2.3 Design supports future adaptation to changes in functional needs	Design of the building parts to support ongoing use in the same or a different design configuration in the same building
3. Ease of recycling	3.1 Parts made of homogenous materials with minimal unnecessary treatments or finishes	Specification of components and constituent parts made of homogenous materials, the same materials or materials mutually compatible with recycling processes. Finishes, coatings, adhesives or additives should not inhibit recycling

(continued)

Table 3.8 (continued)

General deconstruction aspect	Specific design aspect to address	Description (EU 2021c)
	3.2 Constituent materials can be easily separated	Possibility to separate components and parts into their constituent materials
	3.3 There are established recycling options for constituent parts or materials	The part or material is readily recyclable into products with a similar field of application and function, thereby maximising their circular value

lead to outcomes involving the potential for reuse, pure stream recycling, mixed stream recycling, material recovery or energy recovery;

- outline a draft of product and material passports, identifying the elements from design drawings, bill of quantities and materials and fit-out plans;
- "*calculate a circularity score that takes into account the best practical outcome (based on consideration of ease of disassembly and the potential for reuse, recycling and recovery) and the weighted masses (and optionally values) for each end-of-life outcome*" (EU 2021c).

The above-mentioned "circularity score" should be aligned with the waste hierarchy set out in the Waste Framework Directive, namely (in order of best outcome first) (Directive 2008/98/EC): direct reuse; preparing for reuse; pure stream recycling; mixed stream recycling; material recovery; energy recovery; inert or non-hazardous landfill; hazardous waste disposal.

Figure 3.3 provides an illustration of the logic path that should be followed when defining which kind of re-strategy and end-of-use/-life outcome is viable and suitable for each building element, component or material.

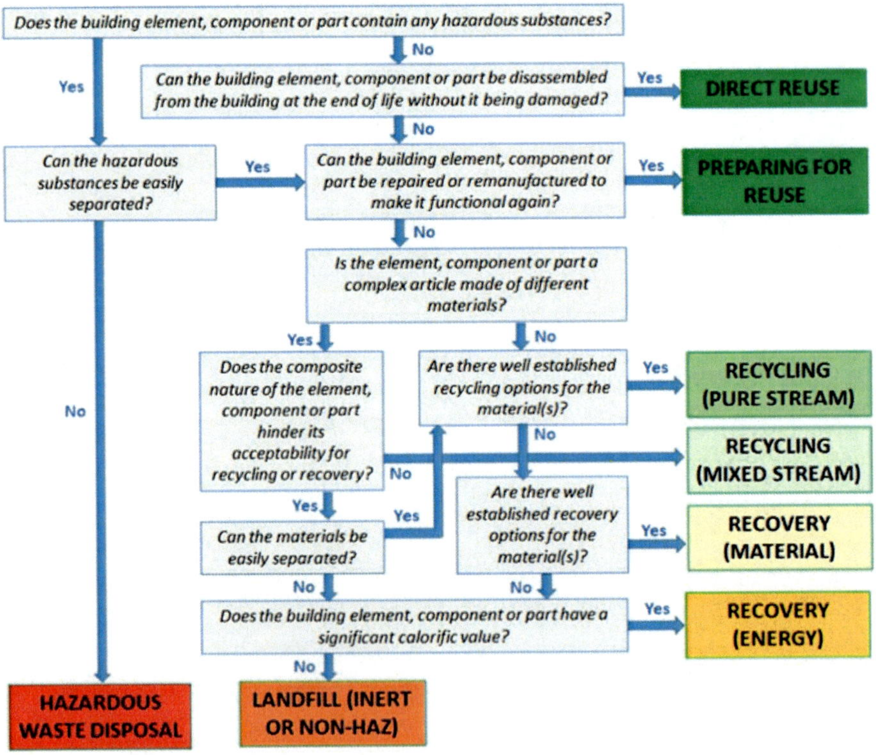

Fig. 3.3 General logic path to follow in ordert to identify the most sustainable and suitable re-strategy applicable to building elements, components, parts or materials. *Source* EU (2021c)

3.2 Existing Support Tools for Circularity in Building Design: Features and Added Value

The present paragraph is aimed at introducing key voluntary tools developed at the European and international level for supporting the integration of circularity principles within current building design practices. In particular, the paragraph focuses on: (i) the international standards ISO 20887 "Sustainability in buildings and civil engineering works. Design for disassembly and adaptability. Principles, requirements and guidance" by the International Organization for Standardization (ISO); and (ii) the "Circular Buildings Toolkit" by ARUP and Ellen MacArthur Foundation (EMF). These tools are deepened in the following sub-paragraphs, highlighting: key features, goals and objectives; structure and principles; involved stakeholders and main targeted audience.

3.2.1 The Contribution of the International Standards ISO 20887 "Sustainability in Buildings and Civil Engineering Works. Design for Disassembly and Adaptability"

The ISO 20887 standard "Sustainability in buildings and civil engineering works. Design for disassembly and adaptability. Principles, requirements and guidance", published in 2020 by the International Organization for Standardization (ISO), provides an overview of Design for Disassembly and Adaptability (DfD/A) principles and strategies for integrating these principles into building and product design processes. Hence, the standard addresses building owners, architects, engineers, product designers and manufacturers, construction companies, maintenance and deconstruction operators with the aim of supporting them in understanding/adopting/implementing the DfD/A principles and requirements (ISO 20887:2020).

As scope of application, the ISO 20887 includes buildings of different destination of use (e.g. exhibition, commercial, industrial and residential) whether they are under design or construction, new construction, under maintenance, renovation or requalification.

The standard is focused on two key concepts for circularity, namely: disassembly and adaptability. The standard proposes the adoption of these concepts in building design as a way to overcome commonly-employed traditional construction processes and assembly methods that do not consider disassembly and deconstruction. By integrating DfD/A concepts early in the concept design phase (when it is still cost-effective to do so with respect to the use phase (MacLeamy 2004; van der Zwaag et al. 2023), it is possible to increase the likelihood that activities during the phases of use, maintenance (including repair, refurbishment, replacement) and end-of-life (including the disassembly, the re-strategies—such as reuse, remanufacturing, recycling—and the disposal) will be conducted more efficiently in terms of time, tools and materials, labor costs, energy. Hence, the design team has a crucial role in integrating DfD/A, thus in: definining the most suitable construction techniques ensuring DfD/A and gathering all the involved supply chain operators—including product suppliers, constructors, facility managers, decommissioning operators—to share the same vision and approaches of DfD/A within the project and guaranteeing that they all have sufficient knowledge and understanding to achieve the expected DfD/A project results. Therefore, to this end, the proper drafting of the Briefing Document (BDs) represents the starting point towards a circular project. In particular, on the basis of these premises, when developing the Briefing Document (BDs), in order to effectively integrate DfD/A principles and requirements, it is fundamental to firstly analyze the reference context conditions to subsequently assess the potential trade-offs in terms of DfD/A approaches and their impacts in terms of life cycle costing and life cycle assessment. In this regard, according to the ISO 20887 standard, it is important to consider the following aspects and characteristics of the project, that could influence the scope and applicability of DfD/A (ISO 20887:2020):

- location physical context, e.g. allowance for change due to economic conditions, demographics, topography, etc.;
- location cultural context, e.g. prescribed methods of construction, labor versus material costs, etc.;
- type of owner, e.g. owner occupant, investor, corporate, developer, government, etc.;
- use type(s) of buildings, e.g. institutional, healthcare, residential, retail, commercial, educational, industrial, warehouse/storage, etc.;
- building typologies, e.g. high-rise, low-rise, detached, etc.;
- construction technologies, e.g. framed column and beam construction, precast concrete, steel-framed building, platform frame construction, load bearing construction, composite construction, curtain wall building, etc.
- construction materials, e.g. concrete, masonry, steel, heavy timber, light-wood framing or their combinations;
- size and dimension, e.g. footprint and height, plot space, space types and allocations, spatial organization, etc.;
- building design life, i.e. proposed life of "first-use" and any anticipated "further use" identified by the client;
- project/building performance goals related to: (i) environmental, social and economic sustainability targets; (ii) construction, function and operation;
- potential climate change effects and hazard zone requirements (e.g. flooding, earthquake, etc.) that could add requirements for enhancing the ease of adaptability or disassembly for major repair;
- construction/deconstruction schedule and timeframes, i.e. time to construct and/or disassemble the construction works;

– external agents, i.e. factors that could influence the degradation or obsolescence giving rise to additional inspections and maintenance interventions and substitutions.

Moreover, the standard introduces an interesting insight concerning the balance between durability and adaptability. Indeed, if on one hand durable elements and solutions ensure the reduction of maintenance and substitution interventions, limiting the environmental impacts over their life cycle, on the other hand, if not properly designed, they could hinder adaptability and replacement, depending on the reversibility degree of their connections with other products or systems (i.e. ease to disassembly). Hence, the "independence" of the durable elements must be accurately estimated based on the component life expectancy (i.e. service life).

In addition to these general considerations, the ISO 20887 standard provides specific DfD/A principles (Table 3.9) articulated into the two key categories of adaptability and disassembly.

3.2.2 "Circular Buildings Toolkit" by ARUP and Ellen MacArthur Foundation for the Design of Circular Buildings

The Circular Buildings Toolkit (CBT), developed by ARUP with the Ellen Macarthur Foundation, has the aim to support designers, construction clients and asset owners to understand how to shift towards circularity by providing a set of common guidelines and circular strategies. The CBT seeks to overcome the linear "take–make–waste" consumption model by adopting circular design principles and management models aimed at reducing waste by keeping products and materials in use for longer. According to ARUP, the toolkit supports building stakeholders in (ARUP 2023a): (i) designing and building for longer term use; (ii) developing a new materials mindset for a reduced use of virgin and non-renewable materials; (iii) learning from completed projects, choosing from a library of completed circular building projects that prove the practicality and value of circularity.

In particular, the CBT proposes a high-level framework[1] articulated into four key broad approaches, namely: (i) build nothing; (ii) build for long term value; (iii) build efficiently; (iv) build with the right materials (ARUP 2023b), which are subsequently parted into sets of prioritized strategies (Fig. 3.4) and then detailed actions (Table 3.10), that support design teams in integrating circular economy principles within new Real Estate projects right from the very preliminary phase of concept design. The framework is also aligned with the latest regulations of the European regulatory framework, including the EU Taxonomy and the EU Level(s).

[1] The Framework of the Circular Buildings Toolkit by ARUP and Ellen Macarthur Foundation is available at: https://ce-toolkit.dhub.arup.com/framework.

Table 3.9 DfD/A principles and requirements according to ISO 20887. *Source* ISO 20887:2020

DfD/A principle	Definition	Advantages
Adaptability principles		
Versatility	Ability to accommodate different functions with minor system changes	– Possibility to easily change the layout through the relocation of components (reconfiguration of the space), without significant modifications – Reduced number of strip-out and fit-out over the building life cycle – Reduced overall building footprint, required floor area, costs, and resources by having one space that accommodates multiple uses – Increased building utilization
Convertibility	Ability to accommodate substantial changes in user needs by making modifications	– Suitability to changing needs, either on an infrequent or irregular basis or at a future point in time by designing the space or fit-up to facilitate minor, non-structural modifications to interior spaces (e.g. partitions, ceiling, and finishes) or furnishings – Convertibility for multiple uses can improve the profitability of a space, as well as reducing the need for other facilities, thereby reducing resource and energy use
Expandability	Ability of a design or the characteristic of a system to accommodate a substantial change that supports or facilitates the addition of new space, features, capabilities and capacities	– Allowance for either vertical or horizontal additions in floor space. Expanding horizontally, the design shall facilitate the disassembly of existing walls, envelope, or partitions so that space can be expanded without significant damage and materials can be re-used, either on the existing project or another

(continued)

Table 3.9 (continued)

DfD/A principle	Definition	Advantages
Disassembly principles		
Ease of access to components and services	Exposed connections are left accessible for disassembly or modification of components, assemblies, or systems within a constructed asset	– Allowance for a material, component, or connector of an assembly, especially those with the shortest anticipated life cycle, to be easily approached, with minimal damage to and impact on it and adjacent assemblies – Reduced replacement time and generation of unnecessary waste during the replacement or maintenance of materials or components – Increased ease of disassembly by making the connections more visible
Independence and reversible connections	Quality that allows parts, components, modules and systems to be removed or upgraded without affecting the performance of connected or adjacent systems	– Optimized disassembly for both re-use and upgrade by maximizing independence of the functional requirements of parts, components, modules and systems – By designing the building systems or "layers" to stand independently it is possible to facilitate the removal, adjustment, replacement, or upgrade of components (that may have a different use-life or useful-life) – The separation of structure from enclosure will greatly facilitate adaptation and disassembly, thus the re-use of systems, spatial adaptability, and functional adaptability – Separating long-lived components from short-lived components will facilitate adaptation and reduce the complexity of disassembly, allowing specific types of materials to be removed one at a time, thus facilitating the collection process for recycling or upgrading

(continued)

Table 3.9 (continued)

DfD/A principle	Definition	Advantages
	Reversible connections can be disconnected and/or disassembled for easy alterations and additions to structures	Connections should: (a) Leave necessary room on all sides to accommodate disassembly options; (b) Require the same standard tools for assembly as well as disassembly; (c) Use universally recognized connection methods that do not damage the materials being connected or the surrounding areas; (d) Minimize interdependency of different materials, products, components or systems: – The use of reversible connections instead of fixed fasteners to connect products or components can allow for easier disassembly. Not only can the material be used again but the connectors (e.g. screws, bolts) can also be re-used – By making products easier to take apart, so that constituent components are not harmed, elements can be re-used directly as long as they meet performance requirements. Materials can also be readily separated by material type and then serve as inputs for other products through recycling processes
Avoidance of unnecessary treatments and finishes	Use of recyclable or reusable materials either on the exterior or in the interior of a constructed asset	– Avoid the use of paint, veneer, or other finishes
Supporting re-use (circular economy) business models	Reusability—Ability of a material, product, component or system to be used in its original form more than once and maintain its value and functional qualities during recovery to accommodate reapplication for the same or any purpose	– This approach is intended to achieve the economic and environmental benefits associated with disassembly. The ultimate reusability depends on the value of the material and extent to which it can retain that value and function after being removed or disassembled

(continued)

Table 3.9 (continued)

DfD/A principle	Definition	Advantages
	Refurbishability—Ability to restore the aesthetic and functional characteristics of a product, building or other constructed asset to a condition suitable for continued use	– The refurbishing of products can reduce the consumption of natural resources – Depending on the intended design life of the construction works, refurbishability can help reduce operating and maintenance costs (the supplier shall make information available on how a product is refurbishable) – The use of construction components that can be refurbished allows for an increase in their service life
	Remanufacturability—Ability of a product to be disassembled and refabricated at the end of its useful life in a manner that provides restoration to a condition suitable for resale	– Remanufacturable products are designed in a manner that allows for complete upgrading: products can be inspected and assembled to their individual elements, and damaged pieces can be repaired or replaced. The product is restored to an "as new" condition for resale by the manufacturer – The use of construction components that revert to the ownership of the original manufacturer (e.g. via take back programs) can reduce waste and lower costs
	Increased recycling—Use of recycled materials, either directly or as feedstock within a manufactured product	– The use of recycled materials reduces the reliance on primary non-renewable materials, costs and environmental burdens – Recycling produces both economic and environmental benefits (e.g. reduced energy, water, and natural resource consumption and reduced emissions) by replacing virgin materials with recycled materials within the life cycle

(continued)

Table 3.9 (continued)

DfD/A principle	Definition	Advantages
	Future recycling/recyclability—Ability of component parts, materials or both to be separated and reprocessed from products and systems and subsequently used as material input for the same or different use or function	– If a material is readily recyclable (i.e. it can be diverted from the waste stream and, through existing processes, facilities, and markets, returned to the economy), a portion of its initial cost can be recovered at the end of its useful life through separation and resale as a recyclable commodity
Simplicity	Quality of an assembly or system that is designed to be straightforward, easy to understand and meet performance requirements with the least amount of customization	– The design principle of "simplicity" reduces the number of elements, components, subcomponents, or materials to the minimum required to execute the intended function thus facilitating repair interventions and also reducing the likelihood of failure or breakdown – The more homogeneous the materials of a structure, the simpler it is to sort materials on site for re-use and recycling. For components, this can help reduce the tools and techniques required for disassembly, thus simplifying the process of sorting on site and making the potential for reprocessing more attractive, due to the larger quantities of the same or similar items

(continued)

Table 3.9 (continued)

DfD/A principle	Definition	Advantages
Standardization	Standardisation involves: a. Dimensions (e.g. standard height and sizes that allow for multiple types of use) b. Components (e.g. standard lengths/spans to facilitate further re-use and ease of replacement) c. Connections (e.g. connecting parts which can be separated using readily available and standard tools) d. Modularity (e.g. volumetric pods which can be slotted together, added to or taken away to promote adaptable living or working environments)	– Standardized parts make it easier for contractors to disassemble structures – Standardized parts can allow for easier transportation, storage, and re-use – Due to the interchangeability of standardized parts and components, standardization facilitates simplicity, adaptability and further re-use in both design and the various phases of constructed assets – Selecting standard-size material can accommodate re-use and upgrading, since materials can be purchased with greater ease (and more cost effectively) when they are of standard dimensions – Standard sizes reduce the creation of on-site off-cut waste for everything from timber, plywood, masonry, and insulation panels to floor tiles – Prefabricated elements or components and a system of mass production reduce site work and allow greater control over component quality and conformity
Safety of disassembly	Accessibility to accurate information on the original materials and assembly methods used for an element, together with details of any subsequent major renovation, supports the correct disassembly sequencing (according to the disassembly plan), facilitating the further re-use and recycling	– The availability of a disassembly plan, thus the easy access to information on assembly methods used for an asset (product, component, subcomponent, etc.), support the safe disassembly, reducing risk of injury and handling difficulties – The disassembly plan, developed according to the adopted reversible design principles, allows the non-destructive disassembly

Fig. 3.4 The Circular Building Toolkit (CBT) by ARUP and Ellen MacArthur Foundation. *Source* ARUP (2023b)

As shown in Fig. 3.4, each of the four approaches is articulated into strategies. Each strategy is then broken-down into practical actions[2]—described in Table 3.10. Moreover, the framework assigns to each strategy a metric (KPI—Key Performance Indicator) with the related unit of measurement to evaluate the actual compliance of the project with the strategy (Table 3.11).

The application of such a tool while developing the Briefing Documents (BDs) for building design could be very useful to: (i) define the most suitable and sustainable design approach to the specific intervention; (ii) guide the design activity toward circularity objectives; (iii) translate circularity objectives into concrete and viable design actions, acting practically on the concept- and then detailed-project; (iv) support the dialogue with all the different involved building stakeholders relying on a common circularity framework; (v) assess and make visible—by means of the proposed metrics—the integration of circular economy principles in the project and the related arising added-value and benefits.

[2] The "detailed actions" proposed for each circular strategy of the Circular Building Toolkit (CBT)—by ARUP and Ellen MacArthur Foundation—are further broken-down into elementary sub-actions, available at: https://ce-toolkit.dhub.arup.com/strategies.

Table 3.10 The "detailed actions" of the circular strategies proposed by the Circular Building Toolkit (CBT) by ARUP and Ellen MacArthur Foundation. *Source* ARUP (2023c)

Structure of the Circular Buildings Toolkit (CBT) by ARUP with the Ellen Macarthur Foundation				
Approach	Strategy		Action	
Build nothing	1	Refuse unnecessary new construction	1.1	Reuse, renovate or repurpose an existing asset
Build for long-term use	2	Increase building utilization	2.1	Increase the multi-use potential of building spaces
			2.2	Create the general physical conditions to enable multi-use implementation
			2.3	Design for an increased utilization of regularly "empty" spaces
			2.4	Design local building performance units so that they can work at various space configurations and requirements
			2.5	Make us of versatile/flexible/movable internal walls for the space layout to support multi-use
	3	Design for longevity	3.1	Design for future climate adaptability/resilience
			3.2	Prioritize standardized, modular elements over bespoke/tailor-made solutions, and avoid complex building geometries
			3.3	Investigate Product-as-a-Service schemes for components expected to have a short or medium service life in the project
			3.4	Maximize the durability of the building structure through careful selection, protection and maintenance of components
			3.5	Ensure the individual service life of envelope systems, components, products and materials align with the minimum service life of the building
			3.6	Make use of whole life-cycle cost assessment (WLCC) as design assessment tool
			3.7	Issue a building materials passport document for the project
	4	Design for adaptability	4.1	Increase convertibility: choose architectural massing, a structural grid and a foundation layout compatible with all likely future uses

(continued)

Table 3.10 (continued)

Structure of the Circular Buildings Toolkit (CBT) by ARUP with the Ellen Macarthur Foundation				
Approach	Strategy		Action	
			4.2	Increase convertibility: allow for changes in building use by designing the building envelope to allow for more than one use, or to allow modifications in window size and spacing
			4.3	Increase convertibility: make passive provision accounting for possible changes to MEP systems, provide a plant replacement strategy that avoids waste
			4.4	Develop and issue an adaptability manual document
	5	Design for disassembly	5.1	Develop reversible connections between the building super-structure elements
			5.2	Allow access to reversible connections between the structure and building services
			5.3	Develop and issue a disassembly manual document for the building
Build efficiently	6	Refuse unnecessary components	6.1	Refuse redundancy in spaces and overestimated headcounts
			6.2	Eliminate/reduce the need for on-site parking space
			6.3	Prioritize passive and simple servicing strategies over overly complex ones
			6.4	Refuse finishes where possible
	7	Increase material efficiency	7.1	Avoid material intensive deep underground and high-rise construction
			7.2	Reduce the material use intensity in the building structure via material-efficient structural forms and techniques, such as hybrid and/or composite solutions
			7.3	Reduce dimensions of the building structure components through selection of high strength materials
			7.4	Use advanced engineering practices to improve material efficiency of structural and envelope components

(continued)

Table 3.10 (continued)

Structure of the Circular Buildings Toolkit (CBT) by ARUP with the Ellen Macarthur Foundation

Approach	Strategy		Action	
			7.5	Reduce material waste at production and construction through off-site prefabrication of the building structure and envelope components
Build with the right materials	8	Reduce the use of virgin and non-renewable materials	8.1	Maximize the use of reclaimed components for all building layers
			8.2	Use concrete with high secondary content
			8.3	Use engineered timber (or other biobased materials) in building structures
			8.4	Use bio-based rapidly renewable materials for the interior design concept
			8.5	Reduce the use of critical raw materials
	9	Reduce the use of carbon-intensive materials	9.1	Track the embodied carbon footprint during design and set an ambitious overall embodied carbon target for the project
			9.2	Track the embodied carbon footprint of building structure and set a target which is below the regionally recommended thresholds
			9.3	Track the embodied carbon footprint of building envelope and set a target which is below the regionally recommended thresholds
			9.4	Track the embodied carbon footprint of building systems and set a target which is below the regionally recommended thresholds
			9.5	Track the embodied carbon footprint of building fit out components and set a target which is below the regionally recommended thresholds
			9.6	Design for digital information management and provide sufficient information for LCA

(continued)

Table 3.10 (continued)

Structure of the Circular Buildings Toolkit (CBT) by ARUP with the Ellen Macarthur Foundation				
Approach	Strategy		Action	
	10	Design out hazardous/pollutant materials	10.1	Track all environmental impacts during design through detailed LCA, not just carbon, and set an ambitious target for the overall project (all layers, including realistic functional and service lives of components)
			10.2	Ensure that building materials and products are not on the 'Living Building Challenge (LBC) Red List'
			10.3	Use on-site electric equipment to reduce the use of fossil fuel driven machines on site, to in turn reduce the impact of nitrogen, smog and particulate matter emissions in the area
			10.4	Avoid the use of hazardous/pollutant materials in the services inside the building
			10.5	Avoid the use of hazardous/pollutant materials in the space
			10.6	Manage hazards of legacy materials in existing buildings

Table 3.11 CBT circular strategies and related KPIs by ARUP and Ellen MacArthur Foundation. *Source* ARUP (2023c)

Strategies and KPIs of the Circular Buildings Toolkit (CBT) by ARUP with the Ellen Macarthur Foundation

Approach	Strategy		KPI
Build nothing	1	Refuse unnecessary new construction	Reuse of existing usable surface Share of reused floor area as percentage of total project gross floor area [%]
Build for long-term use	2	Increase building utilization	Total building utilization Cumulative hours of occupancy, defined as total hours * person spent in the building on a weekly basis, and normalized per square metre [hrs/m^2]
	3	Design for longevity	Value retention and recovery over whole life cycle Life cycle cost (according to EU Level(s) Indicator 6.1 Life cycle costs) [\$/m^2/yr], accounting for real functional service lives of the building and of each individual component, as well as assessing potential returns due to sell-back schemes and high residual value of components
	4	Design for adaptability	Adaptability potential Adaptability score, defined as per EU Level(s) Indicator 2.3. Adaptability, Table 6 (quantitative rating resulting from a qualitative assessment)
	5	Design for disassembly	Disassembly and recovery potential Ease of recovery + ease of reuse and recycling scoring, defined as per EU Level(s) Indicator 2.4 design for deconstruction (assessment methodology based on DGNB TEC1.6 Ease of recovery and recycling)
Build efficiently	6	Refuse unnecessary components	Conceptual material efficiency To account for material use reductions not achieved through technical optimizations, but rather conceptual decisions, a material use intensity factor per functional unit over building life cycle is introduced. The functional unit is to be set depending on the building typology, for example, total material use intensity per workstation/hotel bed/resident, etc. [kg/unit/yr]

(continued)

Table 3.11 (continued)

Strategies and KPIs of the Circular Buildings Toolkit (CBT) by ARUP with the Ellen Macarthur Foundation			
	7	Increase material efficiency	Material use efficiency Total material use intensity by area, and over whole building life cycle, accounting for all building materials [kg/m²/yr]
Build with the right materials	8	Reduce the use of virgin and non-renewable materials	Material Circularity Indicator (MCI) by Ellen MacArthur Foundation
	9	Reduce the use of carbon-intensive materials	Whole life cycle GHG emissions Carbon emissions intensity measured over the whole building life cycle, as defined by Level(s) Indicator 1.2 Life-cycle global warming potential [kgCO_2eq/m²/year]
	10	Design out hazardous/ pollutant materials	Environmental cost Whole life cycle environmental impact cost per floor area, and over the whole life cycle period as defined, as defined by the Dutch MPG methodology [€/m²/year]

3.3 Adding New Contents to Briefing Documents (BDs) for Building Design

The development of Briefing Documents (BDs) is crucial for guiding the building design and integrating circularity principles in the project definition. The BDs set out the vision, design approaches, technical and functional requirements for the building project (ISO 20887:2020), that will have an impact on the subsequent building process phases, till the use and management, and end-of-life.

To include circular design criteria, specifications and requirements in an effective way, the drafting of the BDs should be based on an iterative collaborative dialogue between the client and all the members of the multidisciplinary project team, including designers, architects, engineers, environmental consultants, etc.

The key topics to first discuss in the BDs development phase in order to define the circular strategies and actions to implement in the project are:

1. the definition of the useful-lives and service-lives (to plan) of the building components and systems. The definition of the needed and potential durations of use and life of the building elements represents one of the key starting points to (i) define and attribute to the technical elements the suitable design criteria for allowing and plan disassembly, reuse, remanufacturing or recycling activities; (ii) identify from the design stage the most suitable and sustainable organizational models (e.g. traditional purchase, renting, product-as-a-service, etc.). In particular, the following aspects need to be defined by the client together with the project design team (ISO 20887:2020) in order to support the subsequent definition, planning and programming of circular re-strategies:

   - required service life of the different building systems, sub-systems, elements and components—ranging from temporary to permanent during the whole building useful-life;
   - expected uses of the construction work over its required service life—ranging from (i) single use type to multiple use types (e.g. residential, commercial, office, etc.), and from (ii) single use spaces to multipurpose spaces;
   - ownership of the assets—this issue is twofold: (i) the ownership of the building, for instance the building can be owned for a long-term use or as investment to be rented or leased to multiple tenants; and (ii) the different systems, sub-systems, elements and components that compose the building can be directly bought with a consequent transfer of ownership to the client or they can be rented or bought as-a-service with the ownership (and related responsibility) retained by the manufacturer;
   - review of the Circular Economy regulatory framework to define mandatory project requirements in terms of environmental sustainability and circularity (compliance);
   - review of the voluntary Circular Economy standards and certifications to (i) define concrete circular strategies, detailed actions and step-by-step procedures and to (ii) define ambitious objectives and targets for attaining project

incentives or economic bonuses, for instance starting from the review of national policies for the construction products and material provision and supply, etc.;
- definition of objectives, targets, benchmarks and Key Performance Indicators (KPIs) concerning circularity, including adaptability, ease to disassembly, repairability, etc. (e.g. with the support of Level(s), ISO standards, etc.) and related outcomes in terms of potential for reuse, remanufacturing, recycling, etc. and derived reduction of life cycle impacts;
- definition of the "planned obsolescence", on the basis of the estimation of the probability/predictability of the (functional, technological and economic) obsolescence of the different building systems and elements in order to draft a first program of interventions for disassembly the obsolete elements and make them available for reuse, remanufacturing or recycling;
- design and supply of durable and adaptable building elements. The durability (expressed in terms of useful-life) and the adaptability (e.g. ease-to-disassembly for reuse) of the designed elements must be carefully assessed. Building elements with a high durability are always highly recommended since they require less frequent maintenance, repair and replacement interventions. Moreover, according to the assessment of the length of their use-life and the planned obsolescence, the elements must also be ease to disassembly and reuse (including design principles such as accessibility, independence, simplicity, ease of re-use, remanufacturing and recyclability);
- definition of the possible degradations and profiles of uses (estimation of the entity of the utilization, for instance based on the expected number of occupants/users) of the different building parts, creating a first knowledge base for the development of the Maintenance Plan;
- operation and maintenance planning—including (i) setting of maintenance requirements to ensure that the building and its parts will maintain their functional, economic and aesthetic value overtime; (ii) the definition of who will develop a Maintenance Plan for maintaining overtime the asset based on the indication provided right from the design stage (e.g. Disassembly Plan and design-for-disassembly criteria) and (iii) who will be responsible for documentation storage and transfer of information throughout all the different phases of the building process.

2. The use of materials, components and products that have recycled content must be incentivized starting from the concept design stage. As highlighted by the ISO 20887 standard: "*recycled materials might be available as the result of full or partial demolition, deconstruction, disassembly, strip out and other removal of assets. The application of pre-demolition or pre-refurbishment audits can help to identify possible materials and products to be incorporated into subsequent recycling initiatives. Recycling efficiency is optimized with homogeneous material. Recycled products, or products that contain a proportion of recycled content, might also be available from third-party suppliers*" (ISO 20887:2020).

3. Connected with the first two points, another key topic to discuss in the BDs development pahse is represented by the definition of the supply network (e.g. demolition operators, refurbishers, fit-out companies, local remanufacturers, dealers, resellers, logistic operators, environmental consultants, logistics and distribution operators, waste managers, etc.) needed for the implementation of the circular re-strategies in the use and management phase of the building process till the end-of-life (the first configuration of the supply network should be enriched and updated overtime according to the specific needs). It is important to define the roles and responsibilities of each involved stakeholder and the relationships among them (agreements).
4. In the light of the results and considerations introduced in the previous three points, it is important to draw conclusions by defining general approaches towards a circular design and construction and what to avoid both in the design phase and also in the future in the event of any interventions that may require subsequent planning (e.g. requalification or extraordinary maintenance). In particular, these design intents or "circularity-oriented design stage actions" can be represented in the form of checklists to be developed and shared with all the stakeholders involved in the Building Process. In this regard, an example of a checklist is shown below:

   - Limit the use of cast-in-situ concrete;
   - Limit the use of low-reversibility solutions (e.g. use of resins and adhesives; use of coatings difficult to remove; etc.);
   - Prefer mono-material components;
   - Use durable materials;
   - Use modular elements;
   - Use standardised elements;
   - Use prefabricated elements;
   - Use reversible mechanical and easy-to-access connections;
   - Ensure independence of the parts (the different parts of an element or a system are independent and separable, thus they can be removed without damaging other elements);
   - Ensure the accessibility of building elements and systems (in the use and management stage) for maintenance operations or for disassembly aimed at a circular re-strategy (reuse, remanufacturing, etc.);
   - Size properly the components for manual handling;
   - etc.

5. It is necessary to understand the entity of the "information need" requested to be able to plan and implement reuse practices. In this regard, it is important to:

   - use a classification and coding system for building elements to allow all the stakeholders involved to uniquely identify the elements and to allocate to the elements the related properties, data, information and documents;

- create a digital document archive able to act as a repository of updated documents (e.g. as-built drawings, product certifications and guarantees, contractual documents, etc.). This archive can also be a module of a wider technological information system (e.g. BIM or Real Estate Information Platforms) for the future management of the building;
- draft, during the design phase, a Disassembly Plan (or Deconstruction Plan), including information and step-by-step procedures on the deconstruction/disassembly techniques to apply, useful for favouring circular interventions over time also by different operators, without lose useful information and critical data. In fact, as also recalled by ISO 20887 "*any component, module or system to be disassembled requires a disassembly plan that is considered at the onset of design to ensure its effectiveness*" (ISO 20887:2020);
- development of a Maintenance Plan, including Use Manual (indications and instructions for future users on how-to-use in the best way the building elements), Maintenance Manual (listing the possible degradations, anomalies and obsolescence of building elements and related preventive/corrective interventions—for maintenance operators) and a Maintenance Program (schedule of the preventive interventions to maintain the functional, economic and aesthetic value of the building elements overtime);
- development of Product and Material Passports, i.e. for each material or building element (product, component, etc.) it is necessary to develop a digital sheet according to a common template that include all the relevant data and information for allowing a circular use of the building element itself (e.g. physical and functional features, owner, cost of purchase and repair, number of use-cycles, localization within the building, accessibility and independence of the element, occurred maintenance interventions, disassembly procedures, etc.). The Passport should also link to the other above-mentioned documents (e.g. Maintenance Plan and Disassembly Plan);
- development of a Construction and/or Demolition Waste Management Plan according to the contents of EU Level(s) Indicator 2.2 "Construction and Demolition waste and materials" (see Sect. 3.1.1).

All these documents—which feed the Digital Building Logbook[3] (EU 2020)—must be appropriately inserted in the digital repository of the adopted information

[3] The Digital Building Logbook is "*a common repository for all relevant building data. It facilitates transparency, trust, informed decision making and information sharing within the construction sector, among building owners and occupants, financial institutions and public authorities. A digital building logbook is a dynamic tool that allows a variety of data, information and documents to be recorded, accessed, enriched and organised under specific categories. It represents a record of major events and changes over a building's lifecycle, such as change of ownership, tenure or use, maintenance, refurbishment and other interventions. As such, it can include administrative documents, plans, description of the land, the building and its surrounding, technical systems, traceability and characteristics of construction materials, performance data such as operational energy use, indoor environmental quality, smart building potential and lifecycle emissions, as well as links to building ratings and certificates. As a result, it also enables circularity in the built environment. Some types of data stored in the logbook have a more static nature while others, such*

system (e.g. BIM or Real Estate Information Platforms) and linked to the material/product/system properly coded in an unambiguous way. These documents must be accessible to all the involved stakeholders, acting as an always available and updated information base for the implementation of circular practices overtime. Furthermore, a Document Manager must be appointed. The Document Manager will be responsible for updating (or requesting specific operators to update) and archiving the documents, making them always available for consultation. Moreover, in light of this complex documental set, it is advisable to consider the preparation of training programs for the staff members who will carry out the maintenance and preparation-for-reuse activities (including disassembly).

With reference to the activities described in the five above-introduced points, the Level(s), the ISO 20887 standard and the Circular Buildings Toolkit by ARUP and Ellen Macarthur Foundation represent fundamental guidance and support tools to be used from the initial concept design stage and from the very first dialogue with the client to set a common ground and language for the design activities.

Moreover, it is important to underly the key role of other additional support tools and technologies, namely:

- System for building classification and coding. Such a system is needed to set a unique way to recognize and name the different building parts according to a "hierarchical open structure"[4] (Talamo and Bonanomi 2016), creating a common language to share among stakeholders and with the digital tools for the design and management. Among the commonly recognized systems, the Onmiclass by the Construction Specifications Institute (CSI) represents a virtuous option since it is also widely used to provide a classification structure for databases, software, and Building Information Modeling (BIM) (CSI 2023). Indeed, OmniClass provides a standardized basis for classifying and coding building elements and for organizing, sorting, analyze, storing and retrieving related data throughout the whole building life cycle from conception to reuse or demolition (CSI 2023). In particular, OmniClass is derived by the standard: ISO 12006-2 "Building construction.

as data coming from smart meters and intelligent devices, are dynamic and need to be automatically and regularly updated" (EU 2020).

[4] The hierarchical open structure is *"a tree structure, starting from a level of maximum aggregation (real estate, compound or building) that can be organized in underlying layers, each one having a lower degree of complexity, to the simplest elements, not further decomposable. It is applicable both to spaces and technical elements [...] it allows to: (i) allocate information at different hierarchical levels; (ii) organize and extend the hierarchy to further levels, if necessary (vertical extension); (iii) add new elements at the same hierarchical level, if necessary (horizontal extension); (iv) aggregate information with respect to different hierarchical levels (i.e. for classes of technical elements, typologies of technical elements, typologies of spaces, etc.).*

On the basis of the assumed hierarchical structure, all the items constituting the physical system can be identified by assigning to them a unique code for recognition. The code may be composed in many ways (it can be done manually or automatically generated), but, in general, the code should: (i) lead to the unambiguous identification of each entity (spatial, technical or documentary); (ii) be as "speaking" as possible, i.e. explicit with respect to the type of entity and level of hierarchy organization; (iii) be coherent with the criterion of the assumed hierarchical structure" (Talamo and Bonanomi 2016).

Organization of information about construction works. Part 2: Framework for classification" and it is articulated into 15 hierarchical inter-related tables. Each table deals with a speccific building aspect. A table can be used individually or in combination with other tables to classify, describe aggregated data, link complex information. Specifically:

Table 11—Construction Entities by Function;
Table 12—Construction Entities by Form;
Table 13—Spaces by Function;
Table 14—Spaces by Form;
Table 21—Elements;
Table 22—Work Results;
Table 23—Products;
Table 31—Phases;
Table 32—Services;
Table 33—Disciplines;
Table 34—Organizational Roles;
Table 35—Tools;
Table 36—Information;
Table 41—Materials;
Table 49—Properties.

– Digital building information systems or Information Platforms (e.g. BIM, Real Estate Information Systems, etc.) to collect, store, make available, update, analyze and manage the building knowledge base (including data, information and documents) improving the circularity-related decision-making during the different phases of the building process. Indeed, from the design stage, these tools allow an integrated and horizontal building Information Management. Indeed, the knowledge base is accessible by all the involved stakeholders (even beloging to different phases of the building process), who can—with different authorities—update and enrich the base adding data and information overtime. In this way, it is possible to avoid information loss, which could hinder the implementation of circular practices, and to inform the operators to properly perform their (maintenance, disassembly, remanufacturing, etc.) work during building use phase. By using an Information Platform starting from the design phase it is possible to (Atta 2023):

  - automatically classify and codify technical and spatial elements designed/to be designed (e.g. according to Omniclass);
  - develop first drafts of Material and Product Passports and then complete them in the detailed design and post-construction phases. Storage and update of the Passports, linking them directly to the elements recognized by the coding and classification system (see previous points);
  - create 2D and 3D graphic representations (digital twins) of the building elements (with related physical, dimensional, functional features) editable over time;

- starting from the preliminary design phase, create a single information base (digital database) updated throughout the building life cycle as a repository of data, information and documents (of different nature and concerning different aspects, from design drawings to product certifications, to administrative documents, to Material and Product Passports, Maintenance Plans, Disassembly Plans, etc.);
- simulate re-action scenarios (reuse, remanufacturing, repourposing, recycling, etc.) for assessing the related technical and economic feasibility and for estimating environmental impacts in order to inform the design process;
- define systems of indicators, benchmarks and KPIs for the assessment of circularity in the detailed design phase and calculate them automatically (the system calculates the indicators by sourcing the necessary data from the single central database). The system will then allow—in the subsequent phases—to monitor and control over time performances linked to circularity according to what defined in the design phase.

Concluding, the Briefing Documents (BDs) for building design must be understood as a structured system of informative and preparatory documents for the detailed design which must also take into account the aspects of circularity. This documentary apparatus must not be set aside once passed to the subsequent phases of the Building Process but it must serve as an information base for guidance and orientation also for the construction and management phases. In fact, this document system is enriched with new circularity requirements related to design-for-disassebly, reuse, remanufacuring, repurposing, etc. and new documents (albeit still draft and qualitative in the design phase) such as Material and Product Passports, Maintenance and Disassembly Plans, disassembly instructions, etc. that support the successful implementation of circular approaches.

The BDs must be inserted in the digital document archive of the Information Platform, as a guiding reference throughout the whole Building Process. It is fundamental to always take into consideration the original requests and objectives of circularity even in the in-use phase to inform decision-making overtime. Therefore, it is essential to make the BDs (including the first drafts of Maintenance and Disassembly Plans and the Passports of materials and products) and the consequent detailed project easily accessible (trhough the Information Platform) from all the stakeholders gradually involved in the Building Process.

References

ARUP (2023a) Circular Buildings Toolkit. Discover strategies and actions to develop truly sustainable buildings. https://www.arup.com/services/climate-and-sustainability-services/circular-economy-services/circular-buildings-toolkit. Accessed May 2023

ARUP (2023b) Circular Buildings Toolkit. Framework. https://ce-toolkit.dhub.arup.com/framework. Accessed May 2023

ARUP (2023c) Circular Buildings Toolkit. Strategies/actions. https://ce-toolkit.dhub.arup.com/strategies. Accessed May 2023

Atta N (2023) Remanufacturing towards circularity in the construction sector: the role of digital technologies. In: Arbizzani E et al (eds) Technological imagination in the green and digital transition, CONF.ITECH 2022. The Urban Book Series. Springer, Cham

CSI—Construction Specifications Institute (2023) About Omniclass. https://www.csiresources.org/standards/omniclass/standards-omniclass-about. Accessed May 2023

EU (2020) Definition of the digital building logbook. Report 1 of the study on the development of a European union framework for buildings' digital logbook. https://op.europa.eu/en/publication-detail/-/publication/cacf9ee6-06ba-11eb-a511-01aa75ed71a1/language-en

EU (2021a) JRC technical reports. Level(s) indicator 2.2: construction and demolition waste and materials user manual: Introductory briefing, instructions and guidance (Publication version 1.1). https://susproc.jrc.ec.europa.eu/product-bureau//product-groups/412/documents. Accessed May 2023

EU (2021b) JRC technical reports. Level(s) indicator 2.3: design for adaptability and renovation user manual: Introductory briefing, instructions and guidance (Publication version 1.1). https://susproc.jrc.ec.europa.eu/product-bureau//product-groups/412/documents. Accessed May 2023

EU (2021c) JRC technical reports. Level(s) indicator 2.4: design for deconstruction User manual: Introductory briefing, instructions and guidance (Publication version 2.0). https://susproc.jrc.ec.europa.eu/product-bureau//product-groups/412/documents. Accessed May 2023

EU (2023a) How Level(s) applies to you. https://environment.ec.europa.eu/topics/circular-economy/levels/lets-meet-levels/how-levels-applies-you_en. Accessed May 2023

EU (2023b) How does Level(s) work? https://environment.ec.europa.eu/topics/circular-economy/levels/lets-meet-levels/how-does-levels-work_en. Accessed May 2023

EU (2023c) Level(s) in action. https://environment.ec.europa.eu/topics/circular-economy/levels/lets-meet-levels/levels-action_en. Accessed May 2023

MacLeamy P (2004) MacLeamy Curve (WP-1202). http://www.msaipd.com/MacleamyCurve.pdf. Accessed May 2023

Talamo C, Bonanomi M (2016) Knowledge management and information tools for building maintenance and facility management. Springer

van der Zwaag M, Wang T, Bakker H, van Nederveen S, Schuurman ACB, Bosma D (2023) Evaluating building circularity in the early design phase. Autom Constr 152:104941

Standards and Laws

Directive 2008/98/EC. Directive 2008/98/EC of the European Parliament and of the Council of 19 November 2008 on waste and repealing certain Directives. https://eur-lex.europa.eu/legal-content/EN/TXT/?uri=celex:32008L0098

ISO 20887:2020 Sustainability in buildings and civil engineering works. Design for disassembly and adaptability. Principles, requirements and guidance

Chapter 4
Circular Provision of Building Products and Services: Integration of Invitations to Tender (ITTs)

4.1 New Circular Approaches to the Procurement of Building Products and Services

The EU COM (2008) 400 "Public procurement for a better environment" introduces the Green Public Procurement (GPP), highlighting its potential to push and support public clients to include environmental considerations in their procurement processes and procedures for the provision of goods and services. Already in 2008, the EU identified the construction sector as a priority field. Indeed, in the COM (2008) 400, the EU identified ten "priority" sectors for GPP, selected on the basis of "*the importance of the relevant sector in terms of the scope for environmental improvement; public expenditure; potential impact on the supply side; example setting for private or corporate consumers; political sensitivity; existence of relevant and easy-to-use criteria; market availability and economic efficiency*" (COM 2008: 400). The first identified priority sector is in fact Construction, covering "*raw materials, such as wood, aluminium, steel, concrete, glass as well as construction products, such as windows, wall and floor coverings, heating and cooling equipment, operational and end-of-life aspects of buildings, maintenance services, on-site performance of works contracts*" (COM 2008: 400).

In this regard, the European Commission's Joint Research Centre (JRC) aims at defining sets of common EU GPP criteria and requirements that can be integrated into public procurement procedures to reduce the environmental impact associated to the provision of goods, services or works. At present, the EU has published the updated version of the EU GPP criteria related to some specific vertical fields, e.g. furniture; indoor cleaning services; office building design, construction and management; public space maintenance; etc. (EU 2023a).

In particular, the novel circularity-oriented approaches introduced within the GPP towards a "Circular Procurement" can be summarized as follows (EU 2023b):

N. Atta, *Green Approaches in Building Design and Management Practices*,
Digital Innovations in Architecture, Engineering and Construction,
https://doi.org/10.1007/978-3-031-46760-8_4

- *Multiple life cycles thinking.* Indeed, the Circular Procurement requires an understanding not only of how products are made (e.g. recycled content, environmental impacts of production processes, etc.) and where they come from (e.g. origin of raw materials, supply distance, etc.) but also of how they will be used over time and what will happen to them after their first use cycle and after the end of their useful life (e.g. definition of possible re-strategies—i.e. reuse, remanufacturing, recycling, etc.—according to the EU waste hierarchy);
- *Acting at the very preliminary phase.* The early procurement stages represent the "*greatest opportunities for reducing the environmental impacts and maximising the economic and social impacts within the cycle*" (EU 2023b). Hence, acting right from the phase of development of the ITT—by including requests and metrics related to circularity issues to proactively engage potential suppliers—represents a promising strategy to effectively and efficiently implement circular practices during the product/service provision (contract period);
- *Alternatives to ownership models.* The adoption of innovative service-based models (the client pays for the "use of"/ "access to" the product) characterized by the "extended producer responsibility" where the ownership is retained by the manufacturer—as circular alternative to the traditional single sell/buy model characterized by the transfer of ownership from the seller to the buyer—represents a facilitator to implement reuse and remanufacturing strategies, giving the possibility to the producer to become the remanufacturer that prepare the post-use products for a subsequent use;
- *Collaborative stakeholder network.* Circular procurement and lifecycle thinking require the establishment of a collaborative stakeholder network that cooperate to implement circular approaches to procurement. The network involves a wide set of stakeholders, holders of different interests, roles, know-hows and resources, including investors, decision-makers, project managers, facility managers, manufacturer, providers, dealers, waste managers, environmental consultants, etc. According to the "GPP Training toolkit Module 5: GPP & the Circular Economy" (EU 2023b), the key steps to establish stakeholder networks and collaboration relationships are:

    a. identifying the relevant stakeholders;
    b. mapping their interest and influence;
    c. engaging with stakeholders, e.g. to identify current capabilities and potential to offer alternatives (e.g. circular products, business models, etc.);
    d. communication, e.g. through pre-competitive dialogue, with markets and potential suppliers.

- *Circular clients.* Actions to raise the awareness of clients/users and increase their level of acceptance of reused products and new product-service or product-sharing formulas are essential. Indeed, raising awareness of clients/users about thinking at the long-term consequences and impacts of their purchasing decisions represents almost a precondition for the activation of circular practices.

4.2 Circular Procurement for a Sustainable Building Management

The present paragraph introduces key supporting tools for a Circular Procurement of products and services in the use and management phase of the Building Process. In particular, the paragraph focuses on: (i) the "Circular Economy Procurement Framework" by Ellen MacArthur Foundation (EMF), highlighting possible actions to integrate circularity and circular economy principles within the different steps of the procurement journey; and (ii) the ISO/DIS 59010 "Circular Economy. Guidance on the transition of business models and value networks" standard by the International Organization for Standardization (ISO), analyzing the proposed business-oriented methodology aimed at supporting organizations in accomplishing the transition from linear to circular business models and from value chains to collaborative value networks.

4.2.1 The Contribution of the "Circular Economy Procurement Framework" by Ellen MacArthur Foundation: Integrating Circularity in the Procurement Journey

The term "Circular Procurement" introduced in the context of the GPP is also recalled by the Ellen MacArthur Foundation, which have developed a circular procurement tool for supporting the integration of circularity objectives within procurement processes. In particular, the Circular Economy Procurement Framework by Ellen MacArthur Foundation is articulated according to the key phases of a Procurement Process (Fig. 4.1) and it "*presents the circular intervention points that can be used by an organisation in every step*" (EMF 2020).

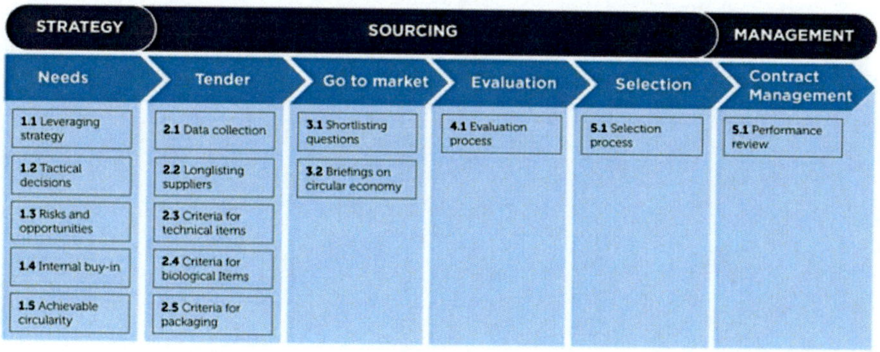

Fig. 4.1 The Procurement Journey by Ellen MacArthur Foundation. *Source* EMF (2020)

The tool is divided according to three macro-phases of the Procurement Journey identified by EMF, namely: strategy, sourcing and management. These macro-phases are then articulated into phases, which are in turn subdivided into steps (Tables 4.1 and 4.2).[1] For each step the aim of the tool is twofold: it provides a checklist useful to assess the level of circularity adoption and/or it proposes strategies for integrating circularity within the step.

The three macro-phases are the following:

1. Strategy. It involves the "Needs" phase and consists in the setting of the circularity objectives for the product/service sourcing and the definition of the sourcing strategy to implement;
2. Sourcing. It includes the following phases: "Tender", "Go to market", "Evaluation" and "Selection". In particular, it entails the definition of circularity-related requests and specifications to include in the Invitation to Tender (ITT), the analysis of the level of maturity concerning the circularity topic of the potential product/service suppliers on the market and the definition of circularity-related criteria for evaluating and selecting the received tenders (offers);
3. Management. It refers to the "Contract Management" phase and it involves the continuous improvement of the product/service provision. The performance both of the provided product/service and of the partnership with supplier/s towards circular economy are constantly monitored, assessed and consequently adjusted in order to ensure a "mutual value generation" (EMF 2020) (win–win relationship).

[1] See the complete "Circular Economy Procurement Framework" by Ellen MacArthur Foundation at https://emf.gitbook.io/circular-procurement/-MB3yM1RMC1i8iNc-VYj/overview (EMF 2020).

Table 4.1 Phases and steps of the Procurement Journey by Ellen MacArthur Foundation. *Source* EMF (2020)

Phase	Phase description	Step		Step description/ Examples (EMF 2020)
Needs	This step involves confirming the sourcing need, validating its objectives with internal stakeholders, and mapping out the related risks and opportunities	1.1	Leveraging strategy	Define the circular economy vision and strategy, related main circularity needs and objectives and assess the ability to fulfil them. The circular economy strategy on a company level may include decisions such as switching from virgin material suppliers to take-back schemes. This would influence the business model, require a gradual approach and time to implement, and would need the involvement of various stakeholders
		1.2	Tactical decisions	Definition of more circular supply options with respect to the traditional ones
		1.3	Risks and opportunities	Definition of risks and opportunities related to the circular economy strategy and tactical decision according to the specific application context (e.g. type of goods or service, supplier, etc.)

(continued)

Table 4.1 (continued)

Phase	Phase description	Step		Step description/ Examples (EMF 2020)
		1.4	Internal buy-in	The value case for adopting a circular economy approach should be clear for all the internal stakeholders, including those parties who will be responsible overtime for the asset or service throughout their life cycle
		1.5	Achievable circularity	Assessment of the achievability and viability of the defined strategies and tactical decisions (although complete circularity is the target, the maturity and capability of the supply base may not allow this from the start)
Tender	This step includes defining the tender criteria, analysing the market and long-listing the suppliers	2.1	Data collection	When including circular economy criteria in the tender process, it is important to refer to the defined strategic circular economy objectives to create a general framework about criteria to be met and the related needed data to collect
		2.2	Longlisting suppliers	Consider circular economy aspects when longlisting the suppliers
		2.3	Criteria for technical items	Circular economy criteria for the technical cycle
		2.4	Criteria for biological items	Circular economy criteria for the biological cycle

(continued)

Table 4.1 (continued)

Phase	Phase description	Step		Step description/ Examples (EMF 2020)
		2.5	Criteria for packaging	Circular economy criteria for packaging
Go to market	This step includes shortlisting the suppliers and then executing and managing the tender process	3.1	Shortlisting questions	Circular economy questions to consider for pre-qualification questionnaire
		3.2	Briefings on circular economy	Organize a circular economy focused pre-tender briefing
Evaluation	This step involves evaluating the responses to the tender and clarifying the proposals	4.1	Evaluation process	Circular economy questions to consider for evaluation
Selection	This step is about selecting the supplier by focusing on value creation opportunities	5.1	Selection process	Circular economy questions to consider for selection
Contract manager	This stage is about the ongoing supplier performance review and management to ensure mutual value generation	6.1	Performance review	Examples of circular economy KPIs

Table 4.2 Checklist for circularity integration within the steps of the Procurement Journey by Ellen MacArthur Foundation. *Source* EMF (2020)

Procurement step		EMF checklist towards circularity integration (EMF 2020)	
1.1	Leveraging strategy	Q.1.1.1	Can you leverage the company's circular economy strategy (if you have one) to make procurement decisions more circular? (Can you define what full circularity looks like for your company? How does it influence your key sourcing needs?)
		Q.1.1.2	Can you reframe the need and find circular economy opportunities?
		Q.1.1.3	Can you think of alternative sourcing opportunities by finding ways to substitute the virgin inputs?
		Q.1.1.4	At what point do you need to engage other internal and external stakeholders to help make these decisions?
1.2	Tactical decisions	Q.1.2.1	Can you explore the opportunities to reuse as-is or repurpose internally?
		Q.1.2.2	Can you choose non-ownership-based sourcing options?
		Q.1.2.3	Can you embed circular economy criteria in your requirements?
		Q.1.2.4	What is your optimal supply chain structure to address your circular economy needs?
		Q.1.2.5	What due diligence do you need to conduct on circular economy products?

(continued)

Table 4.2 (continued)

Procurement step		EMF checklist towards circularity integration (EMF 2020)	
		Q.1.2.6	Can you choose the payment arrangement that enables circularity: • Fixed period (e.g. price/month): potential circularity, but no incentive to limit consumption/ use rates • Pay-per-use (e.g. price/wash cycle) or Pay-per-outcome (e.g. price/ provision of light or price/ service of floor covering): greater chance for circularity, as the supplier is incentivised to provide the service with minimum consumption of resources
1.3	Risks and opportunities	Q.1.3.1	Can you consider the following aspects in relation to your sourcing need: • technical aspects • compliance culture • sourcing locations • supply chain capability/capacity • the need to develop an after-market
		Q.1.3.2	What are the circular economy related risks and opportunities addressed by the sourcing process in question?
1.4	Internal buy-in	Q.1.4.1	Can you consider who needs to be involved in the conversations at the strategic and operational levels?
1.5	Achievable circularity	Q.1.5.1	Can you agree internally how to manage and assess a potential partial circular model—with a view to supporting suppliers in their transition to a fully circular model?

(continued)

Table 4.2 (continued)

Procurement step		EMF checklist towards circularity integration (EMF 2020)	
2.1	Data collection	Q.2.1.1	Can you define your circular economy criteria with the following aspects in mind: • It helps to have criteria that are measurable, objective, transparent and verifiable • When you communicate them to your suppliers, consider how to allow for fair competition • You may need to pay special attention to SMEs and the development of their capacity to respond to such criteria • Pay attention to how far up the supply chain it is necessary to go for adequate fulfilment • Allow suppliers to challenge the criteria if they can see opportunities to improve circularity
2.2	Longlisting suppliers	Q.2.2.1	What is the present capability of the supply market to meet or exceed your sourcing needs and the required circular economy criteria?
		Q.2.2.2	How early can you engage your potential suppliers to understand their intent and preparedness to meet your circular requirements?
		Q.2.2.3	Can you consider the following aspects? • New technologies, alternative goods or services and new business models • Existing supplier capabilities • Available reverse logistics structures • Circular economy maturity of the market • Existence of local suppliers and/or partners with local economic impact • Certification system like ISCC to certify the whole chain of custody • The impact extraction, production, and transport has on climate

(continued)

Table 4.2 (continued)

Procurement step		EMF checklist towards circularity integration (EMF 2020)	
2.3	Criteria for technical items	Q.2.3.1	Can you purchase items that are used more? • Can you purchase through business models that increase utilisation (e.g. supplier can provide repair, reuse, rental and remanufacturing options at scale)? • Can you choose items that are designed, created and manufactured to be durable, repaired or refurbished so that they align with a business model that keeps them at their highest value? • Can you make sure that all items that are made and purchased will be used?
		Q.2.3.2	Can you purchase items that are made to be made again? • Is there a system in place to collect and return these items for reuse, repurpose, refurbishment, remanufacturing or recycling, thus making sure they don't end up as waste? • Can you purchase items that use packaging made from reusable, recyclable or compostable materials?

(continued)

Table 4.2 (continued)

Procurement step		EMF checklist towards circularity integration (EMF 2020)	
		Q.2.3.3	Can you purchase items that are made from safe and renewable inputs? • Can you purchase items that are free from hazardous chemicals, and thus respect the health of ecosystems? • Can you purchase items the production of which (including chemicals used during manufacturing and finishing processes) is fully decoupled from the consumption of finite, non-renewable resources? • Can you purchase items made from post-consumer recycled content (where technically possible) both to decouple from finite feedstocks and to stimulate demand for collection and recycling? • Can you purchase items which, if (partially) made from virgin inputs, use inputs from renewable feedstocks, where proven to be environmentally beneficial, and, where relevant, are sourced from regenerative sources? • Can you purchase items that are manufactured, distributed, sorted and recycled using renewable energy? • Can you purchase items that, through their production, maximise resource efficiency (water, energy, material use etc.)?
2.4	Criteria for biological items	Q.2.4.1	Can you source/purchase ingredients that are grown regeneratively?
		Q.2.4.2	Can you source/purchase ingredients that are made from by-products of other processes?

(continued)

Table 4.2 (continued)

Procurement step		EMF checklist towards circularity integration (EMF 2020)	
		Q.2.4.3	Can you utilise the entire value of the ingredients you purchase? Do you have a strategy/plan in place to valorise by-products of the ingredients you purchase?
		Q.2.4.4	Can you source/purchase ingredients that are grown locally where appropriate?
		Q.2.4.5	Can you source/purchase diverse and/or seasonal ingredients?
2.5	Criteria for packaging	Q.2.5.1	Can you eliminate problematic or unnecessary packaging through redesign and innovation, getting rid of materials, components or formats that: • are not reusable, recyclable or compostable • can be avoided altogether • hinder or disrupt recycling • have a high likelihood of being littered • or contain hazardous chemicals?
		Q.2.5.2	Can you replace single-use packaging with reusable formats, i.e. refillable or returnable, or choose to implement alternative delivery models?
		Q.2.5.3	Can you purchase packaging or plastics that are 100% recyclable or compostable, meaning they are effectively recycled or composted in practice and at scale?
		Q.2.5.4	Can you purchase packaging or plastics with recycled content or sourced from renewable (or bio-based) feedstocks?
3.1	Shortlisting questions	Q.3.1.1	What are the required capabilities your supplier must have to fulfil your purchasing needs with the circular economy criteria? If they are not demonstrable at the point of tender, how do they plan on building these in-line with your requirements?

(continued)

Table 4.2 (continued)

Procurement step		EMF checklist towards circularity integration (EMF 2020)	
		Q.3.1.2	What parts of the supply chain can the supplier cover and what parts of potential upstream activities do they have influence over in terms of circular economy requirements?
		Q.3.1.3	How can you help suppliers develop new circular economy capabilities?
		Q.3.1.4	Does the supplier have a good understanding of circular economy principles? Can they articulate them and demonstrate their understanding properly through their activities and offerings? Does the supplier have a sustainability officer and a circular economy strategy? (It's likely that the circular economy approach will be embedded in their sustainability strategy)
		Q.3.1.5	Does the supplier have a Circulytics score (EMF tool that measures companies' circular economy performance, available at: https://ellenmacarthurfoundation.org/resources/circulytics/resources)? If not, would they be willing to complete it?
		Q.3.1.6	Is there a relevant ISO (e.g. 20400) they are certified in?
3.2	Briefings on circular economy	Q.3.2.1	Can you conduct the supplier briefings, setting out the requirements and communicating circular economy opportunities?
		Q.3.2.2	Can you communicate and confirm the selection criteria and objectives during these briefings?
		Q.3.2.3	Can you ensure the suppliers fully understand the commercial and circular economy expectations?
		Q.3.2.4	Can you seek opportunities to gather more circular economy information and potential opportunities to collaborate with your supplier?

(continued)

Table 4.2 (continued)

Procurement step		EMF checklist towards circularity integration (EMF 2020)	
		Q.3.2.5	Can you encourage joint proposals and collaboration among your suppliers with complementary offerings to address the capabilities you need?
4.1	Evaluation process	Q.4.1.1	Can you conduct supplier clarification workshops to ensure they can deliver and meet your circular economy criteria?
		Q.4.1.2	Can you conduct supplier site visits or other verification activities if necessary to see the circular economy elements of a supply chain if possible?
		Q.4.1.3	Can you create and circulate evaluation templates for all key stakeholders to score suppliers?
		Q.4.1.4	Can you run debriefs for suppliers upon disqualification to help them improve their circular economy offerings?
5.1	Selection process	Q.5.1.1	Can you combine the total cost of ownership and circular economy related value in one analysis to maximise value?
		Q.5.1.2	How can you create a negotiation environment which fosters innovation and problem solving—two essential ingredients for a circular economy?
		Q.5.1.3	How can you ensure that all parties in the negotiation have the best possible understanding of the circular economy and available solutions being sought so that the outcome is mutually beneficial?

(continued)

Table 4.2 (continued)

Procurement step		EMF checklist towards circularity integration (EMF 2020)	
		Q.5.1.4	Can you conduct trade-off and sensitivity simulations in order to understand different scenarios and the circular economy value versus the up-front monetary cost?
6.1	Performance review	Q.6.1.1	How can you create an open communication stream with your supplier to periodically evaluate how well they fulfil your circular economy needs?

The tool introduces some key circularity-related topics, which can be further applied to the specific field of building management (see Sect. 4.3).

In particular:

- *New product and service supply models based on as-a-service formulas.* "Usage-based models" or "consumption-based pricing models" are based on a service provision and payment scheme where the user pays according to the use of resources. The user accesses to the product (product-as-a-service) and pays: for the time of use of the product (pay-per-use), for a predefined timeframe of use of the product (pay-per-period) or for the performance received from the product (pay-per-performance). The payment amount is therefore dynamic and varies according to usage. This allows a single product to be used multiple times even by different users individually or simultaneously (sharing). This consumption-based model can bring benefits in terms of reducing overprovisioning and waste (maximizing product usage). Furthermore, the profit of the provider is based on reuse and it will be higher the more the product is used and reused. Periodically, depending on the needs of the product, the provider will have to carry out maintenance and/or remanufacturing actions on the product in order to restore the original performance and, therefore, start other consequential use cycles. On the other side, the client will always have access to a high-performance product, paying for it only when at use (avoiding incurring high purchase costs).
- *New payment methods (pay-per-)* in the context of service- and use-based models require new capabilities of consumption real-time monitoring and, therefore, the provision of a digital information system/platform and a network of sensors for real-time monitoring of key parameters related to product/service use, including a smart billing system. These processes must be standardized in their procedures and managed automatically by the system, within which each element is identified and recognized by a unique code (already assigned in the design phases, see Chap. 3) that allows the allocation of data/information/documents to the specific building element/equipment/furniture.

- *Fundamental role of data collection and management and Chain of Custody* for ensuring the feasibility of re-strategies. The chain of custody, as defined by the ISO 22095 international standard, is the "*process by which inputs and outputs and associated information are transferred, monitored and controlled as they move through each step in the relevant supply chain*" (BS ISO 22095:2020). Indeed, the chain of custody consists of structured processes and standardized procedures for tracking, verifying and archiving data about the history of the product, reconstructing the entire supply chain in order to guarantee the sustainability credentials of the final product. The product is tracked through each stage of the supply chain, from raw material sourcing, manufacturing, processing, transportation to the sale. These processes are implemented with the support of digital technologies such as Blockchain.[2] The latter can provide real-time access to updated and reliable chain-of-custody information. This means that authorized parties can access chain of custody records at any time, viewing product information, thus simplifying product traceability and management.
- *Need for new circular assessment tools and new metrics* for measuring circularity in order to: (i) insert new evaluation criteria related to circularity issues in the ITTs and, thus, be able to assign measurable targets and performance indicators (KPIs) to these criteria in order to choose the most suitable offer and provider; (ii) evaluate the circularity performance of the product/service delivery during the contract period, monitoring circularity performance (key parameters) over time.

4.2.2 Circular Models for Procurement: The Contribution of the ISO/DIS 59010 "Circular Economy. Guidance on the Transition of Business Models and Value Networks"

The international standard ISO/DIS 59010 "Circular Economy. Guidance on the transition of business models and value networks" by the International Organization for Standardization (ISO) aims to support stakeholders (small-medium-large companies, organizations, etc.) in accomplishing the transition of their business models and value networks from linear to circular, providing a business-oriented methodology to integrate circularity in traditional business scenarios (ISO/DIS 59010:2023).

The standard is articulated into five main sections (Fig. 4.2), representing the fundamental steps for implementing circularity in business organizational models. In particular, Table 4.3 describes the steps, highlighting key objectives and outcomes.

[2] "*Blockchain is literally a chain of blocks, although, in this case, it refers to information (the 'block') that is digitally 'chained' together and often stored in a public database. [...] Blockchain is a digital platform that facilitates transactions by recording and verifying data across different industries in a secure manner. Its decentralized cryptography-based solutions minimize the agency of third-party providers, thereby reducing transaction costs and enhancing transparency*" (Oclarino 2020).

Table 4.3 Procedural steps for integrating circularity within traditional business models according to the ISO/DIS 59010 standard. *Source* Adapted from ISO/DIS 59010:2023

ISO/DIS 59010 Clause		Step		Objective	Outcome
4	Set the transition's goals and its scope	4.1	Define the goals of the transition	Definition of the goals of the transition of the organization to a circular economy business model (e.g. reductions in resource inflows, energy, water, carbon footprint etc.) and consequent setting the related scope and purposes of the circularity performance measurement	List of circular transition goals with related possible qualitative or quantitative indicators/targets/ benchmarks to assess the future results
		4.2	Understand the organization's current business model and value network	Understand the current business model of the organization by investigating the key business elements	Report on current: value proposition; key activities; key resources; customer segment; customer relationships; channels; costs; revenue streams; key partners; externalities
		4.3	Map the value chain and networks of flows	Mapping of existing relationships with other organizations and stakeholders along the value chain, also highlighting: features of the involved stakeholders, roles, activities, material flows, business models, etc.	Diagram of the value chain, graphically representing the relationships and interactions among the involved operators

(continued)

Table 4.3 (continued)

ISO/DIS 59010 Clause		Step		Objective	Outcome
		4.4	Set the scope for addressing circularity	Setting of the scope, boundary and level of ambition, including the definition of: (i) all relevant activities, locations, types of solutions and value networks; (ii) portion of the economic system within which the organization operates (value chain) and over which it can drive the circular flows of resources	Defined scope boundary and sphere of influence
		4.5	Understand the level of circular maturity of the organization	Definition of the level of maturity to refine the scope and boundaries within which the circular actions evolve	Report on the level of maturity on circular practices of the organization
		4.6	Understand current circular performance	Assuming Key Performance Indicators (KPIs) to assess the circularity performance of the current business model and value network within the scope boundaries	KPIs calculation results and analysis report, including indicators related to resource inflows and outflows, energy and water consumptions, economic indicators, etc.
		4.7	Actions that contribute to a circular economy	Identification and assessment of possible actions to re-orient current processes, solutions, or value creation models towards circularity	Report on the actions that: (i) create added value; (ii) contribute to value retention; (iii) contribute to value recovery; (iv) rebuild lost values; (v) support transition to a circular economy

(continued)

Table 4.3 (continued)

ISO/DIS 59010 Clause		Step		Objective	Outcome
5	Determine circular economy strategy	5.1	Identify gaps and opportunities	Identification of possible gaps and opportunities in pursuing the predefined circular goals, on the base of the features, context and level of circular economy maturity of the organization	Report of the gaps & opportunities analysis
		5.2	Determine circular economy strategy based on circular economy principles	Definition of a circular economy strategy from a long-term and sustainable perspective, taking into consideration: the resource availability and the organization ability to secure long term resource accessibility and traceability	Report on circular economy strategy, including: (i) identification of the desired goals; (ii) definition of the value that the organization should create and provide and how the latter should be shared via its business model
		5.3	Address economic rationalization	Analysis of the circularity-oriented business model/s to assess their ability to realize a profit while taking into account the associated risks	Report on economic rationalization, including the estimation of business "cost" and "revenue" elements of the identified circularity-oriented business model/s
6	Transition an organization's business model toward circularity	6.1	Develop a plan	Planning of the activities to implement to shift from linear to circular business model/s. The organization can approach the circular economy strategy by breaking it down into specific step-by-step actions to plan for implementation	Transition Plan that includes schedule of activities to implement, related objectives, needed resources and interim results assessment

(continued)

Table 4.3 (continued)

ISO/DIS 59010 Clause		Step		Objective	Outcome
		6.2	Review the elements of the business model in light of desired changes	The key elements of the traditional business models are reviewed according to the results of the analyses of the previous points, including new circularity integration points and identification of potential risks and opportunities	Reviewed version of the report on: value proposition; key activities; key resources; customer segment; customer relationships; channels; costs; revenue streams; key partners; externalities
7	Transform a value network towards circularity	7.1	Establish shared objectives, strategy and plan	Value network members should jointly investigate the possibility to exploit synergies within the value network and develop together a circular economy strategy based shared objectives and a related common Value Network Transition Plan	Common and shared Value Network Transition Plan, i.e. plan agreed by all the members of the value chain. The Plan should define common and synergistic actions to implement, scheduling of the actions, involved operators and resources, interim expected results and assessments
		7.2	Determine appropriate value network governance	Definition of the governance structure of the value network, pointing out roles, authorities and responsibilities of each member of the value network	Diagram (graphic representation) of the value network configuration, together with a report that states the features of the governance structure in order to ensure a clear allocation of roles, responsibilities and governance authorities to the different involved parties

(continued)

Table 4.3 (continued)

ISO/DIS 59010 Clause		Step		Objective	Outcome
		7.3	Leverage or establish shared infrastructures (physical and digital)	Collaboration between members of a circular value network require shared infrastructures, including: (i) digital information system (or platform) for: data and information exchange, digital material or product passport storage and updating, blockchain app for product traceability and chain of custody, etc.; (ii) shared facilities and assets for implementing the re-strategies, e.g. logistics, warehouses, machinery for work on products (e.g. for cleaning, cut, repair, etc.); (iii) shared knowledge, know-how and skills innovations	Definition and implementation of shared infrastructures for stakeholder communication and collaboration, as well as sharing of: (i) facilities and assets; (ii) know-how and skills
8	Review and monitor for continuous improvement	8.1	Measure and report circular performance	Definition of a system of indicators to assess the circular performance, by measuring the results of the implemented actions	Define and apply a set of relevant circularity KPIs and related targets. Development of periodic performance report, including the results of the measurements and calculations of the circularity indicators as well as a summary of the circularity performance of the implemented actions

(continued)

Table 4.3 (continued)

ISO/DIS 59010 Clause		Step		Objective	Outcome
		8.2	Improve the business model of the organization	Periodic review of the plan, as described in 6.1, according to the results of the circularity performance assessment, also including proposals to improve: (i) the achieved performance; (ii) business model patterns and elements; (iii) access to finance and funding	Reviewed Transition Plan of the organization, including the definition of actions to adjust and enhance the business model/s in a continuous improvement perspective
		8.3	Improve the value network	Review of the common value network strategic plan, as described in 7.1, according to the results of the circularity performance assessment and in the light of: (i) the achieved performance and (ii) the changes of the individual Transition Plan of each involved operator; also considering possible review of the scope of the actions; the relationships of members; and the governance performance and structure	Reviewed Value Network Transition Plan, agreed by all the members of the value chain; also highlighting actions to improve circularity performance and value network configurations and contributions

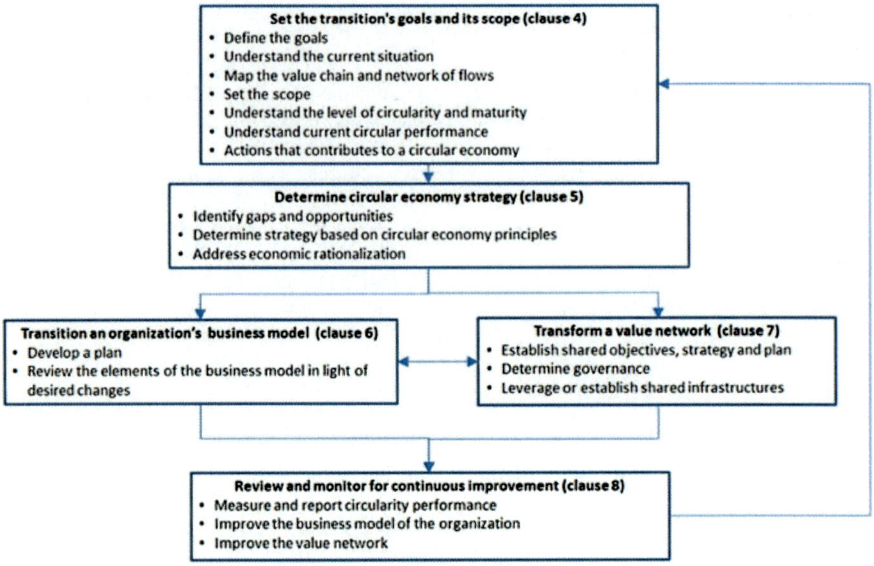

Fig. 4.2 Structure and contents of the ISO/DIS 59010 standard. *Source* ISO/DIS 59010:2023

Based on Table 4.3, it is possible to outline some key considerations that can be applied to the phase of ITT development for the supply of products and services in the context of building Facility Management (FM) activities (see Sect. 4.3). Specifically:

- *Implementation of Product-Service-System (PSS) models.* The PSS model is based on the decoupling of value from the delivered physical product. In fact, this model focuses on the function of the product (service) instead of the sale of the product (ownership). Unlike traditional sale models based on the maximization of the number of sold products, in PSS models the providers are more motivated to develop durable products (lasting for long lifespans). Hence, in addition to environmental benefits and cost savings for users, PSS models promote design for durability and preventive maintenance strategies to extend the product lifespan, since the revenue grows as usage increases throughout the product lifecycle.
- *Importance of the creation of a multi-stakeholder collaborative network* (Fig. 4.3) to synergistically implement circular practices. Indeed, "*expanding the boundary of existing value chains, beyond solely commercial relationships, can create opportunities to develop circular flows of resources by regenerating, retaining or adding to their value. Value networks can take many forms of innovative collaborations and relationships. Organizations within a value network can be multi-sectoral and multi-stakeholder and their relationships can be characterised by new exchanges of materials, financial resources, information, knowledge, technology* etc." (ISO/DIS 59010:2023).

- *Gradual approach to the definition of innovative circular business models.* In particular, the key elements of a business models are firstly recognized and then integrated with circularity aspects (Table 4.4) according to an incremental innovation approach.

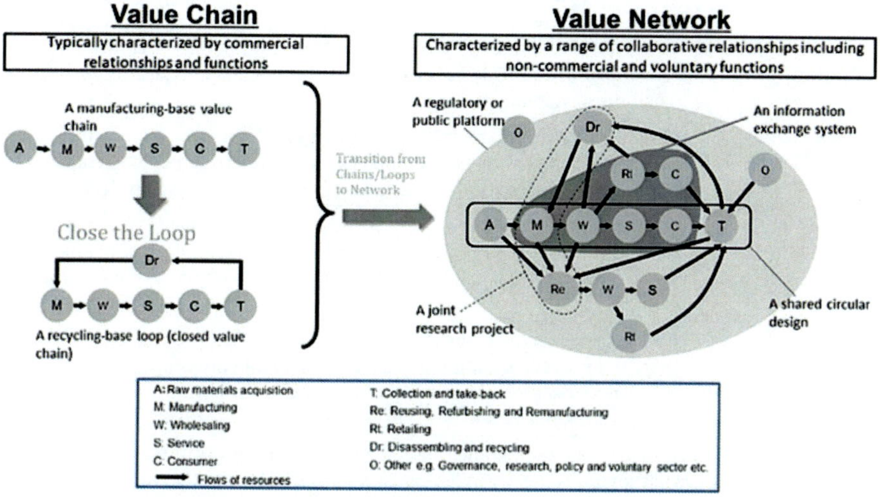

Fig. 4.3 Value Network towards circularity by ISO/DIS 59010. *Source* ISO/DIS 59010:2023

Table 4.4 Integrating circularity within key business model elements. *Source* ISO/DIS 59010:2023

Business model element	Integration of circularity *Source* ISO/DIS 59010:2023
Value proposition	It is essential in the circular economy to extend the engagement and scope of activities such as communication, customer support, customer service, and reverse logistic provision, all value-generating activities around a product or service—often referred to as a Product-Service-System (PSS). A circular value proposition through a product or a service can offer significant differentiation that makes it easier to raise funding and financing for transition, development and scaling
Key activities	Activities focused on improving circularity performance can include process management improvement, equipment modification, technology changes, sharing and virtualization, and improving the design of the product
Key resources	Resources can be explored such as better-performing materials, virtualized materials, resources that enable regeneration and restoration of natural capital, resources that can be obtained from customers, suppliers or third parties
Customer segment	Customer segments are directly linked with value proposition elements, where the value proposition design depicts the fit between value proposition and customer segments. Transitioning to a circular business model can increase attractiveness to new customers
Customer relationships	Building and maintaining relationships can engage customers to suggest what solutions the organization could offer. Additionally, a switch to recycling, reuse or remanufacturing can enhance social-marketing strategies and leverage relationships with community partners. The improvement of the circularity performance can strengthen the relationship between an organization's customers who are committed to social-environmental issues. It may also require increasing the organization's transparency and its willingness to share its circularity performance with customers. Product as a service model or a product take-back system can create long-term links to customers
Channels	Communication with customers on some specific business model features, such as using regenerated materials, secondary raw materials, second-hand components, or remanufacturing, can also be considered if the identified opportunity motivates customers' appreciation for the circular products
Costs	Changes made to each business element will be reflected in the cost structure. The cost structure is usually discussed when the potential benefits of the circular economy are addressed. It can pertain to cost savings related to, for example, the organization's Product Service System (PSS), i.e., its reverse material flow, production costs, costs of product development, or investments
Revenue streams	The revenue streams will be affected by changes in the value proposition. In addition, new pathways to gain revenue from a more circular business model can be explored

(continued)

Table 4.4 (continued)

Business model element	Integration of circularity *Source* ISO/DIS 59010:2023
Key partners	Cooperative networks allow businesses to benefit from suppliers and support an organization in its research, product design, marketing, office support, supply routes, financial functions, production processes, and management. Thus, collaboration enhances essential resource procurement and performance of key activities. For instance, third parties that specialise in industrial waste collection and material recovery may conduct off-site recycling. Outputs or feedstocks, emerging from this processing may then be sold to other industries. Collaborative production, based on the cooperation in the production value chain, allows materials to circulate in a closed material loop
Externalities and enablers	The business environment provided by externalities and enablers is constantly changing. There can be new or amended regulations, radical technology development, disruptions in social infrastructures, or the appearance of new finance schemes—all can provide opportunities for achieving circularity

4.3 Integrating ITTs: New Circularity Requirements for Building Products and Services Procurement

From the review of the support tools for Circular Procurement conducted in the previous paragraphs, it is possible to identify some recurring "lessons learned". These invariants, which can be transferred to the specific context of the outsourcing (ISO 37500: 2014) of building management and Facility Management (FM) services (BS EN ISO 41014:2020), mainly refer to: (i) new models based on the concept of "servitization" (Fig. 4.4), namely Product-as-a-Service (PaaS)—characterized by the access to the use of the product (service)—and Product-Service-System (PSS)—characterized by the delivery of a product functionality and related support services; (ii) new stakeholder network configuration based on partnerships and collaboration; (iii) new circularity requirements for products and services, as well as for manufacturers and service providers; (iv) new metrics for assessing circularity and environmental performance; (v) new digital tools to support data management and performance monitoring overtime.

In regard to these novelties, it is possible to propose some innovative scenarios of integration of current ITTs for the procurement of goods and services in the context of building Facility Management (FM).

In particular, an Invitation to Tender (ITT) can be defined as a formal invitation issued by a company to potential providers to request the submission of proposals for the supply of goods or services (Talamo and Atta 2018). The ITT describes the scope of work, all the specifications and requirements for the supply, the contract timeline, the requested service levels and the related KPIs to monitor overtime (during the contract period) the performance of the provider, as well as the evaluation criteria that will be used to select the winning tender offer, i.e. the provider that is best able to meet the needs of the client (best value for money) (Talamo and Atta 2018).

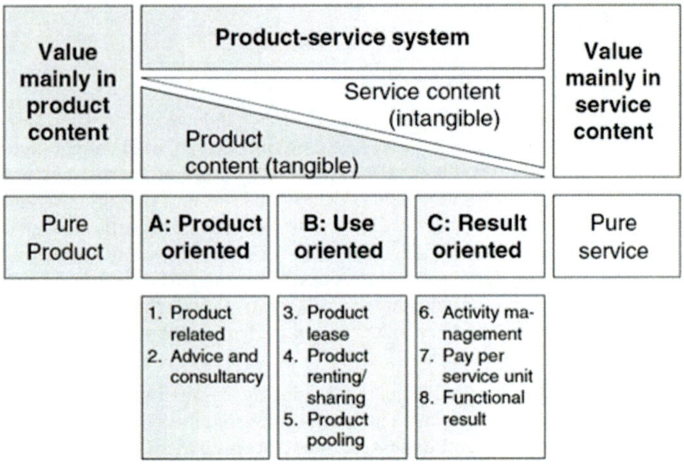

Fig. 4.4 Categories of product-service systems by Tukker (2004). *Source* Tukker (2004)

Hence, the ITT represents a key input to integrate circularity objectives within the supply of goods and services at different levels (product, process, building, company, etc.). In particular, by including circularity requests in ITTs (Table 4.5) it is possible to engage the potential providers from the very early stage of the procurement, assessing their level of awareness and maturity with respect to circular practices. In particular, in Table 4.5 it is proposed a framework for integrating circularity within ITTs articulated in significant reading keys that summarize the main ITTs contents, namely:

a. Supply objectives and scope of work;
b. Required products and services;
c. Organizational models for the supply;
d. Command centre;
e. Building inventory process;
f. Registry system;
g. Information system and ICT infrastructure;
h. Document management;
i. Quality policies;
j. Provider requirements;
k. Procedures for services monitoring and control;
l. Payment methods;
m. SLA and KPI;
n. Penalty system;
o. Incentive system;
p. Responsibility and ownership of information;
q. Information and training;
r. Reporting;
s. Feedback information;
t. Risk management;
u. Reference regulatory framework.

Table 4.5 Proposal of guidelines to support clients in integrating circularity requests within ITTs for the provision of building products and FM services during the building use and management phase

Reading key	Circularity integration points	Level		
		Company	Process	Product
Supply objectives and scope of work	The ITT should firstly define the general circularity goals and objectives with respect to the building product/service provision, referring also to the national, European and international regulatory framework. For instance, the "Procurement Hierarchy"—based on the EU Waste Hierarchy: prevention, preparing for reuse reduce, recycling, recovery, disposal (2008/98/EC Directive)—must be recalled in the ITT, defining the required circular strategies and closed-loop contracts; thus specifying for example the following general circular approaches to pursue during the product/service contract period: – to avoid buying new products, preferring to buy a set of services for the life-extension of products; – to implement preventive maintenance strategies, e.g. condition-based and predictive maintenance to extend the useful life of products; – to prefer the substitution of just a component instead of the replacement of the whole product or system when possible; – to choose products that are durable, ease to be dis/assembled, with high propension to be used for multiple subsequent cycles and with a product certification of sustainability; – to supply products/spare parts and develop services that are always aligned and in compliance with the most updated regulations and directives on Circular Economy (national and European level).	X	X	X
Required products and services	The description of the different required building products and services (e.g. building maintenance, cleaning, energy management, document management, inventory process, space management, waste management, etc.) should be complemented by the related description of circularity objectives and targets (qualitatively or, better, quantitatively expressed—referring also to national, European and/or international standards and regulations). The request of products and/or services must also be integrated with the circular requirements deriving from the previous building design stage, concerning by way of example: – supply of products—e.g. for maintenance or replacement interventions—that are aligned with the circular approaches that guided the choice of the original products (e.g. content of recycled matter, high disassemblability, high maintainability, long lifespan, propensity to reuse, etc.). In this regard, the Briefing Documents (BDs) of the design stage represent a useful and necessary knowledge base on which laying the choices in the building management phase; – the updating of the Product Passports set in the design stage and the request for the development of new Product Passports for new supplied products/spare parts; – the installation and maintenance of the products must always be aligned and consistent with what is indicated within the Disassembly Plan and Maintenance Plan firstly developed in the building design stage. The client should also request to the provider to update overtime all the Product Passports during the use and management phase of the building; – etc. Moreover, depending by the needed service activities, the Client can request: – that the Provider submits a strip-out or pre-demolition audit for the material and product mapping for defining the quantities of waste and of reusable products/materials, the characterization and performance assessment of the latter and the related re-strategies to implement for reuse; – the implementation of a system to monitor and quantify periodically the potential waste arising from the building use and management phase, as well as the materials and products to recycle and prepare for reuse. The Client must always be able to access to these data through the digital information system. All these points need to be explored within the ITT, dedicating a separate specific in-depth section (chapter) to each of them.		X	X

(continued)

Table 4.5 (continued)

Reading key	Circularity integration points	Level		
		Company	Process	Product
Organizational models for the supply	The Client must express its requests with respect to the circular models to implement for the products/services provision. In particular, the Client can: – request for specific organizational models, describing the involved actors, type of supply, adopted re-strategy/ies, payment methods, control system, etc. This is the case in which the Client already owns know-how and experience in the field of circular practices and business models; – request to the Provider to propose for each category of product/service the related organization models for the supply. The proposal should be reviewed by the Client and then agreed by both the parties. This is the case in which the Client does not own the knowledge and experience required to outline the circular organizational models and it exploit the know-how of the Provider. With respect to service-based models, it is important to specify—for type of product/service—if these are: – product-oriented, i.e. the business models is based on the traditional sale of the products but with some extra additional support services; – use-oriented, i.e. with the Provider extended responsibility (ownership of the product is of the Provider) and possibility to product sharing; – result-oriented, i.e. the Client and the Provider only agree on a service result (Service Level Agreement—SLA). For each identified model, the product and/or service payment methods (e.g. fixed fee, pay-per-use, pay-per-period, pay-per-performance, etc.) must be specified in the ITT by the Client or requested to the Provider for a subsequent agreement between the two contracting parties in the negotiation phase. It is important to underline that the shift to the new circular business models should be gradual (trials in the first run-in phase at the starting of the contract period).	X	X	

(continued)

Table 4.5 (continued)

Reading key	Circularity integration points	Level		
		Company	Process	Product
Command centre	The Command Centre (CC) is the "*structure responsible for the planning and coordination of interventions, the monitoring of the outcomes and the management of information flows. This structure, crucial for the process efficiency and effectiveness […] represents a key condition in order to pursue simultaneously at least two basic objectives of the management services: firstly, the client monitoring and control of the compliance with the contract requirements; secondly, the effective integration of the different services included in the contract*" (Talamo and Bonanomi 2016) In this regard, the Client should clearly state in the ITT the functions, the roles and the responsibilities of the Client and the Provider with respect to the definition of the CC. An analysis of the strategies to implement towards the achievement of circularity objective is needed from the very early steps of the tendering process. Indeed, it is fundamental that the Client is aware of its purposes and intentions with regards to all the strategical decisions that the CC will adopt during the service delivery. The result of this analysis must be clearly state in the ITT in order to share with the potential Providers the same strategic vision. Moreover, with respect to circularity objectives, it is important to include in the CC at least one expert on Circular Economy topics (i.e. "Circularity Expert" or "Circularity Manager"). The Client must specify what qualifications the Circularity Expert (or Circularity Manager) must have and what role, responsibility and decision-making authority it has on the circularity issues. This actor can be: – within the Client. When the Client has within its staff an expert figure prepared on circularity issues and with the needed management skills, the "Circularity Manager" or "Circularity Expert" can be covered by a figure internal to the Client. The Client will then ask the Provider to appoint a representative on these issues who will interface with the Client's Circularity Manager in order to share visions, objectives and strategies and to define control activities; – within the Provider. When the Client does not have a Circularity Expert within its staff (or when the Client prefers to focus only on its core business), then it can request the Provider to propose a Circularity Expert from its staff to be part of the CC with the aim of defining strategies and making decisions useful for effective and efficient implementation of circular practices during the contract period. The Client will have to support the activities of the Provider's Circularity Expert with a Client representative. The latter has also the aim of controlling the Provider's Circularity Expert performance and of sharing strategies defined by the Circularity Expert with the Client, who has decision-making authority; – within a Third Party. When the Client and the Provider do not have a Circularity Expert on their staff, or when the Client does not want to give too much control or decision-making authority to the Provider, it is possible to appoint an external subject as a consultant to the Client and the Provider on circularity issues. Whatever the choice of the Client, it will have to explain its requests clearly in the ITT in order to allow the Provider to organize itself accordingly and propose a suitable technical and economic offer (tender).	X	X	

(continued)

Table 4.5 (continued)

Reading key	Circularity integration points	Level		
		Company	Process	Product
Building inventory process	The building inventory is the gradual process of collection of information (including validation and updating) overtime concerning the different aspects of the building (e.g. dimensional, technical, administrative, financial, regulatory, etc.). The Client must specify in the ITT that this process must also be aimed at the production and updating of Product Passports. Therefore, the Client can: – specify in the ITT which information the Product Passports must contain, also providing the related template; or alternatively – ask the Provider to propose a Product Passport template which will then be validated by the Client itself in the pre-contractual phase. The development of the building inventory must be grounded on and aligned with the Registry System.		X	
Registry system	The Registry System represents the system of classification and coding of the Real Estate according to an open hierarchical structure (product breakdown structure—PBS) expandable overtime aimed at identifying in a unique way all the building parts. The Client should specify the classification and coding system adopted in the design stage, e.g. referring to a standard such as OnmiClass. Moreover, the Client should ask to the Provider to adopt the same systems for classification and coding of the building elements in order to enrich overtime the existing knowledge base. The Client should also specify that the Provider must develop and update the Product Passports, Disassembly Plans, Maintenance Plan, Waste Management Plan and other circularity-related documents referring always to the same system of coding (in this way the elements are uniquely identified and it is possible to unambiguously attribute and allocate data, information, procedures, processes and documents overtime to specific assets). By way of example: the classification system and unique identification codes are attributed till the level of detail of the last element that cannot be further disassembled. These codes are therefore also attributed to the sub-components of a system or a product. The codes must be appropriately cited in the Disassembly Plan to effectively guide the operator in the disassembly process.		X	

(continued)

Table 4.5 (continued)

Reading key	Circularity integration points	Level		
		Company	Process	Product
Information system and ICT infrastructure	The Client must clarify within the ITT that the existing information system or information platform for the Real Estate management must be enriched with a specific "Circularity" module able to gather and display relevant data and information and to offer key support service for circularity. An information system for Real Estate management can be defined as a *"decisional and operational support tool, consisting of databases, procedures and functions to collect, store, process, use and update the information necessary for the setting, the implementation and management of the maintenance service"* (UNI 10951:2001). It has a modular and expandable structure consisting of a multiplicity of "modules" relying on the same unique common database. The modules (e.g. space management module, asset management module, operation and maintenance management module, energy management module, risk management module, project management module, etc.) are mutually independent and they can be implemented or not according to the specific needs of the FM service. They can also be progressively acquired over time according to the development and evolution of the organization needs. According to specific procedures for data collection, extraction, storage and processing, the modules are all fed by the central database and, at the same time, they all feed the central database with feedback data coming from the service activities (bidirectional information flow). With respect to each module, the main key functions of the information system involve the development and updating of the building registry, the monitoring of the service activities and the related performance, and the collection and processing of feedback information at the operational level to feed and update the central database. The Client should express in the ITT its requests concerning the implementation of an additional module dedicated to circularity. This new module will be connected to the central databases and to the other existing modules and it will offer functionalities and data-access supporting circular practices, including for instance: – collect, store, link, process and display key data and information useful to support circular practices at different levels, e.g. operational, tactical and strategic; – perform document management activities, including the storage and display of maintenance plans, disassembly plans, site waste plans, performance reports, sustainability reports, etc – implement a monitoring system for ensuring the traceability of products and their performance overtime; – implementation of a circularity-related system of SLAs and KPIs to monitor and assess overtime the circularity performance of processes and products; – collect data useful to develop the so-called chain of custody of the different building parts; – develop digital twins of key (e.g. environmentally critical) building parts to perform simulations of re-strategies in order to define the most viable or suitable re-strategy to implement (e.g. reuse, remanufacturing, repurposing, etc.) also assessing the ease-to-disassembly of the involved elements; – implement a smart billing system for managing innovative as-a-service models based on pay-per-use, pay-per-period or pay-per-performance formulas; – creation of a digital application acting as a virtual marketplace for buying/selling materials and products to reuse/reused; – etc. These requests should be included in the ITT by the Client in order to receive first drafts of the circularity module and its features, to be then discussed and adjusted by both the parties. Due to its novelty, it is advisable to follow a principle of gradualism in order to implement incrementally the module starting from basic functions (to test and validate) up to reach more complex applications.		X	X

(continued)

Table 4.5 (continued)

Reading key	Circularity integration points	Level		
		Company	Process	Product
Document management	The Client must describe in the ITT the activities related to document management and allocate the related responsibilities to the contracting parties (its own staff and staff of the Provider). The client must specify the different documents to manage (e.g. disassembly plan, maintenance plan, product passports, product traceability reports, performance reports, etc.). The Client should also specify: (i) which documents are already available from the previous building process stage, e.g. detailed design and as-built, or the previous management Provider. These documents must guide the work of the new service Provider and they must be updated overtime (in alignment with the original client's intents); (ii) which documents must be developed from scratch and updated by the Provider overtime. The document management activities should be performed using the digital information system or platform (see the previous point), which also acts as document storage, i.e. digital archive.		X	
Provider requirements	In order to ensure a satisfactory service provision, the Client should specify the requirements that the Provider must meet in order to propose a tender. Among the possible Provider requirements, it is possible to mention by way of example: – proved competencies in CVs (e.g. master, professional course, etc.) related to the topics of circular economy, circularity, circular re-strategies, and environmental sustainability in the construction sector; – proved experience (e.g. references) in circular practices within FM services provision; – successful implementation of maintenance plans, disassembly plans, and/or demolition and site waste management plans, etc. for the reduction of waste generation in building practices; – no. years of experience in sustainability subjects within product and service provision for building management; – etc.	X		
SLAs and KPIs system for services monitoring and control	The Client should define the SLAs and the related KPIs concerning circularity targets for the product and service provision. If the Client owns the necessary skills, it should define the circularity-related SLAs and KPIs and specify them within the ITT. Alternatively, if the Client does not have the skills and knowledge required for the proper definition of the circularity-related SLAs and KPIs, it can exploit the knowledge and know-how of the potential Providers. Hence, the Client should clearly request to the Providers in the ITT to formulate a proposal of a set of circularity-related SLAs and corresponding KPIs. In this way the Client can revise the proposed SLAs and KPIs an then jointly agree them with the selected Provider during the pre-contractual negotiation phase in which the two contracting parties can discuss and make an agreement in this regard. For the definition of the circularity-related SLAs and KPIs it is fundamental to refer to the minimum requirements set by law and the targets set by the voluntary European regulatory framework. In relation to the SLAs and KPIs system, penalties and/or incentives can be included to promote the high-quality performance of the Provider.		X	X
Reporting	In the ITT the Client should request to the Provider to periodically (e.g. monthly) draft a performance report that provides a summary of the performance of the service delivery described according to the SLAs and KPIs system. The report should include the calculation of the KPIs and the related results in terms of SLAs satisfaction. The Client should clearly specify the contents of the report, as well as the frequency and methods of the report delivery. The reports act as assessment tools to control the performance and accordingly adjust the strategies towards circularity objectives.		X	

(continued)

Table 4.5 (continued)

Reading key	Circularity integration points	Level		
		Company	Process	Product
Feedback information	With respect to circularity issues, the process of acquisition of feedback information–both at the strategic level (e.g. from report and from users satisfaction surveys) and at the operational level (e.g. from the operators after the execution of interventions), must be specified by the Client highlighting in the 'ITT: – main objectives of the collection and analysis of feedback information at the strategic and operational levels (e.g. identification of possible trends in the implementation of re-strategies, advanced analyses on the re-usability of products based on historical data, assessment of the satisfaction of the innovative as-a-service models, optimization of decision-making processes for product/service provision and building management, etc.); – which feedback information is useful to collect (e.g. number of use-cycles, used/needed spare parts, duration and costs of remanufacturing or repurposing activities, etc.), thus who are the information holders; – which methods of communication and sharing of feedback information (e.g. using a dedicated smart interface in the digital information system or platform, etc.).		X	
Responsibility and ownership of information	With respect to the data, information and documents collected and generated by the Circularity module of the information system during the FM service provision, the Client should define and specify in the ITT: – who will be responsible for the accuracy and reliability of data and information during the service contract period; – that the Client will be the owner of all the data, information and documents collected and generated by the service Provider during the contract period; – that the Client will own the developed knowledge base (i.e. database of the information system and contents of the Circularity module) at the end of the contract period, also specifying the way in which the knowledge base will be handed over to the client and the related deadlines.		X	

(continued)

Table 4.5 (continued)

Reading key	Circularity integration points	Level		
		Company	Process	Product
Information and training	Given the novelty of circular economy and circularity topics within the building management field, the Information and Training in the context of FM service outsourcing represent a fundamental function with respect to the effectiveness and the success of the service provision. In particular, the development and implementation of an Information and Communication Plan allow to: – support information sharing between the Client and the Service Provider concerning relevant topics of interest, including circularity; – share with the external staff the circularity-related objectives and strategies, as well as the related processes and procedures at the strategic, tactical and operational levels; – ensure that external stakeholders are aware of their specific roles, responsibilities and requirements with respect to circularity objectives and tasks; – gain feedback and/or suggestions on circular approaches and practices in the context of the service provision. The development and implementation of a Training Plan allow to: – provide to the external personnel the needed training and educational support to improve their competences and knowledge on circular economy and circularity topics; – mitigate the risk due to lack of knowledge and/or skills on circularity issues. The Information and Training Plans can be developed and implemented by the Client, by the Provider or by a Third Party. In any case, the Client must specify in the ITT who will be responsible for developing the plans and delivering the training sessions to the staff. In particular, it is important to specify or request at least: (i) topics, schedule, duration and place of information sessions and training sessions; (ii) typology of staff (according to their role) that is required to attend the information sessions and the training lessons; (iii) needed tools and software (if any), etc.	X	X	
Risk management	In accordance with the standard BS 8572 "Procurement of facility-related services. Code of Practice", it is important that the Client takes into account the risks arising from the adoption of novel circular approaches, strategies, organizational models and processes. A risk analysis must be conducted before starting the contract period and it has to be periodically reviewed during the service provision. The activities of risk identification and assessment could be performed jointly by the Client and the Provider to "*determine the practical implications of managing innovation and transformation in service delivery against the anticipated benefits*" (BS 8572:2018) and to define tolerance and error ranges of the defined targets and SLAs. Indeed, as suggested by the BS 8572 standard, due to the novelty of the circularity application within building management practices it is advisable to "*ensure that appropriate provisions are incorporated in the service level agreement to accommodate such an arrangement and the changes that might be necessary to the associated service specification*" (BS 8572:2018)		X	X

(continued)

Table 4.5 (continued)

Reading key	Circularity integration points	Level		
		Company	Process	Product
Reference regulatory framework	The Client must request to the Provider to identify the regulatory framework in force on the subject of circular economy and circularity for the building sector and the related requirements to meet. The Client can also list a non-exhaustive set of regulatory tools to be enriched by the Provider in order to define the requirements by law, hence the mandatory specifications, targets and objectives to be in compliance with. Moreover, the Client can also request to the Provider to include voluntary standards and policies useful to reach higher circularity goals. For instance (non-exhaustive list): *EU communications* – COM/2014/398 "Towards a circular economy: A zero waste programme for Europe" – COM/2019/640 "The European Green Deal" – COM/2020/98 "A new Circular Economy Action Plan. For a cleaner and more competitive Europe" – COM/2020/408 "Recovery and Resilience Facility" – COM/2021/573 "New European Bauhaus Beautiful, Sustainable, Together" *EU regulations and directives* – 2008/98/EC Directive (Waste Directive) – 2014/24/EU Directive (Green Public Procurement) – 2018/851 Directive (EPR–Extended Producer Responsibility) – 2019/2088 Regulation (SFDR—Sustainable Finance Disclosure Regulation) – 2020/852 Regulation (EU Taxonomy) – 2022/2464 Directive (CSRD—Corporate Sustainability Reporting Directive) – 2022/1288 Regulation (Do No Significant Harm) *Standards and guidelines at the European level* – ISO 8887-1:2017 Technical product documentation. Design for manufacturing, assembling, disassembling and end-of-life processing. Part 1: General concepts and requirements – ISO 8887-2:2023 Technical product documentation. Design for manufacturing, assembling, disassembling and end-of-life processing. Part 2: Vocabulary – BS 8887-3: 2018 Design for manufacture, assembly, disassembly and end-of-life processing (MADE)—Part 3: Guide to choosing an appropriate end-of-life design strategy – BS 8887-220: 2010 Design for manufacture, assembly, disassembly and end-of-life processing (MADE)—220: The process of remanufacture. Specification – BS 8887-240: 2011 Design for manufacture, assembly, disassembly and end-of-life processing (MADE)—240: Reconditioning – ISO 20887: 2020 Sustainability in buildings and civil engineering works. Design for disassembly and adaptability. Principles, requirements and guidance – ISO/DIS 59020: 2023 Circular economy. Measuring and assessing circularity – ISO/DIS 59010 Circular Economy. Guidance on the transition of business models and value networks – Level(s) European framework for sustainable buildings (2019—in progress)	X	X	X

For each reading key, relevant circularity integration points are proposed, also highlighting the level addressed by the integration, namely:

- company level, i.e. the circularity integration involves a change in the internal structure of the organization (of client and/or provider), also in relation to its work management, role and responsibilities of its internal staff and related skills;
- process level, i.e. the circularity integration implies a change or a necessary addition to current (information, documental, procedural) management processes;
- product/service level, i.e. the circularity integration addresses a change related to the physical product or the intangible service and their supply (e.g. new design approaches, new circularity requirements, etc.).

The wealth of contents and issues, as well as the multiplicity of know-hows and skills characterizing the circular FM supply network—highlighted in Table 4.5—can imply an unprecedented degree of complexity that must be appropriately managed to ensure the achievement of the expected circularity objectives. Indeed, each stakeholder represents a "bearer" of specific knowledge, skills, experiences and competences (e.g. design-for, production, service provision, environmental certification, financial, etc.) that contribute to concretize the circular practice working in synergy with the other network players.

In particular, the network approach to the management of the supply of building products and services for FM and the multiplicity of traditional and novel stakeholders—the latter belonging to the "circularity field" e.g. environmental consultants, dealers, remanufacturers, waste managers, etc.—generates new "interface areas", i.e. areas of expertise that overlap with respect to the different know-how of the involved stakeholders. These overlapping areas create the need for new cross-sectoral profiles of competences and skills required to the traditional building stakeholders in order to guarantee: (i) a good communication along the supply network based on a common understanding of circular approaches (e.g. circular business and organizational models, circular re-strategies implementation, etc.), thus (ii) a good quality and performance of the service provision and the achievement of circularity objectives. In particular, the spreading of circular practices in the building design, construction and management fields may depend in the next future on the presence of professionals with the specific knowledge and skills necessary for the implementation and application of circular service-based models based on reuse, remanufacturing and/or recycling. Hence, the training of traditional building stakeholders, with particular reference to the operators of the FM field, on subjects related to circularity (e.g. circularity regulatory framework, environmental labels and certifications, design-for -disassembly and -reuse, service-based business models, supply network for circularity, etc.) is crucial for the achievement of sustainability and circularity goals within building management and service provision practices, in compliance with the EU regulations and UN SDGs. In this regard, Table 4.6 proposes a set of circularity-related competency profiles for building managers and facility managers, articulated in three increasing levels of maturity.

Table 4.6 Maturity Levels of Circularity-related competency profiles for building and facility managers

		Maturity Level 1	Maturity Level 2	Maturity Level 3
	Proficiency keywords	– Circular re-strategies – Design for modularity/ disassembly/reuse/ recovery – Design for Multiple Use Cycles – Production of construction materials and products – Maintenance and management of construction products – Construction materials and products Market Analysis	*Proficiency keywords of Level 1 and:* – Circular products and manufacturing – Circularity Regulatory Framework – Circularity voluntary standards – Environmental certification – Support documents for circularity (e.g. maintenance and disassembly manuals and plans, product passports, etc.) – Circular Business models (service-based)	*Proficiency keywords of Level 2 and:* – KPIs and metrics for assessing the circularity of practices – Digital marketplace for reused/remanufactured products – Information Platforms for providing circularity support service – Stakeholder network collaboration – User engagement
Category of competence	Knowledge	– Knowledge of basic notions regarding environmental sustainability, waste management and circularity – Notion on waste hierarchy and circular re-strategies (reuse, remanufacturing, reuse, repurposing, reconditioning, recycling, etc.), including related definitions by international standards/regulations – Notions concerning Design-for-modularity/ disassembly/reuse/recovery – Notions about production processes of construction materials and products – Notions about maintenance and management processes of construction products, components and systems – Notions about construction products market (e.g. contract types, warranties, selling methods, payments, etc.)	*All notions listed in Level 1 and:* – Notions about circular production and processes of building materials, products, components and systems – Notions about national & international environmental regulations, certifications and declarations – Notions regarding the contents and development of Product Passports – Notions concerning the contents and development of Disassembly manuals and plans – Notions concerning the contents and development of Maintenance manuals and plans – Notions about product data collection and related tools for the development of support documents for circularity – Notion on the concept of "servitization" and on service-based business and organizational models (e.g. product-service systems, product-as-a-service, etc.)	*All notions listed in Level 2 and:* – Notions regarding Circularity Assessment Tools (metrics, KPIs, indexes) – Notions about information platforms for the creation of digital marketplaces for reused/ remanufactured products – Knowledge of procedure and tools for service/ product monitoring techniques – Notions about information platforms for the supply of circularity support services, including resource/ waste mapping, document management, etc – Notions about Stakeholder Mapping and Review – Notions about collaboration and co-working techniques and tools for stakeholder network alliance – Notion about for techniques for users' active involvement in product/service/model co-design and delivery processes (e.g. surveys, living labs, co-working activities, focus groups, etc.)

(continued)

Table 4.6 (continued)

		Maturity Level 1	Maturity Level 2	Maturity Level 3
	Abilities	– Ability to analyse the feasibility of implementation of the re-strategies (reuse, remanufacturing, reuse, repurposing, reconditioning, recycling, etc.) according to the specificity of the application context (e.g. type of product, profile of use, availability of local remanufacturers, local market context, etc.) – Ability to perform a market review on specific products and geographical scope, including the collection of data and information concerning costs, permanence of products on the market, contract types, warranties, selling and payment methods, etc. – Ability to recognize and assess Design-for-modularity/disassembly/reuse/recovery strategies implemented in the building/product design stage and to adopt in the building use and management stage – Ability to identify the most suitable maintenance strategy according to the reference context (e.g. client priorities, criticality of the building element, normative requests, economic budget, profile of use and degradation of the element, etc.)	*All abilities listed in Level 1 and:* – Ability to review and outline the national & international circularity regulation framework, including environmental regulations, directives, national laws, technical standards, etc – Ability to set, draft and update a Product Passport, collecting and linking data and information from multiple sources – Ability to outline dis/assembly procedural steps and to develop disassembly manuals and plans, also reviewing manufacturer technical datasheet – Ability to identify the most suitable maintenance strategies for specific building elements and to develop maintenance manuals and plans – Ability to assess the feasibility and efficiency of different supply methods for specific building elements and consequent identification of the most suitable circular service-based organizational models (e.g. product-service systems, product-as-a-service, etc.) for each category of building elements – Ability to choose/develop concurrently the circular product (market supply/design), the service and the circular business model – Ability to define the circular business models in relation to the products and/or services and also to: the logistics and distribution processes; localization of remanufacturers in the adjacent geographic area; localization of warehouses for product storage (if needed) – Ability to customize and adjust overtime the service to the users' profiles of use and needs	*All abilities listed in Level 2 and:* – Ability to develop/manage the ICT infrastructure fundamental to effectively and efficiently manage processes and information for activating circular reuse/remanufacturing practices – Ability to operate the integration of heterogeneous information coming from multiple sources within the database of the digital platform – Ability to use real-time data to measure KPIs and performing internal/external benchmarking – Ability to use the ICT infrastructure (e.g. information platform and sensors network) for monitoring in real-time key parameters and to extract useful data form the real-time updated database (query, filter and select data and formats) – Ability to perform different techniques of data analytics (classification, clustering, descriptive/prescriptive analyses) using the digital information platform – Ability to use the modules (expandable overtime) of the digital information platform for the supply of online circularity support services (including modules for resource/waste mapping, document management, re-strategies simulations, etc.) – Ability to develop a website acting as a digital marketplace for reused/remanufactured products – Ability to identify relevant stakeholders for the implementation of circular practices and to map and manage the relationships among them – Ability to engage relevant stakeholders and to promote collaboration through both the use of the common digital information platform (sharing of data, information, experiences, etc.) and the organization of focus groups, roundtables and co-design activities – Ability to engage the users in the circular practices (e.g. incentives or rewards to: promote the give-back of post-use products; share products and services with other users; etc.)

References

EMF—Ellen MacArthur Foundation (2020) Circular economy procurement framework. https://emf.gitbook.io/circular-procurement/-MB3yM1RMC1i8iNc-VYj/overview. Accessed June 2023

EU (2023a) Green public procurement criteria and requirements. https://green-business.ec.europa.eu/green-public-procurement/gpp-criteria-and-requirements_en. Accessed June 2023

EU (2023b) GPP training toolkit module 5: GPP and the circular economy. https://circabc.europa.eu/ui/group/44278090-3fae-4515-bcc2-44fd57c1d0d1/library/61c83b9e-4436-457a-bf40-65e176cba04d/details. Accessed June 2023

Oclarino R (2020) Blockchain's technology of trust. ISO focus September/October 2020, pp 46–53. ISO (ISSN 2226–1095). https://www.iso.org/files/live/sites/isoorg/files/news/magazine/ISOfocus%20(2013-NOW)/en/2020/ISOfocus_142/ISOfocus_142_en.pdf

Talamo C, Atta N (2018) Invitations to tender for facility management services: process mapping, service specifications and innovative scenarios. Springer

Talamo C, Bonanomi M (2016) Knowledge management and information tools for building maintenance and facility management. Springer

Tukker A (2004) Eight types of product-service system: eight ways to sustainability? Experiences from SusProNet. Bus Strateg Environ 13(4):246–260

Standards and Laws

BS 8572:2018 Procurement of facility-related services. Code of practice

BS EN ISO 41014:2020 Facility management. Development of facility management strategy

BS ISO 22095:2020 Chain of custody. General terminology and models

COM (2008) 400. Communication from the Commission to the European Parliament, the Council, the European Economic and Social Committee and the Committee of the Regions. Public procurement for a better environment. https://eur-lex.europa.eu/LexUriServ/LexUriServ.do?uri=COM:2008:0400:FIN:EN:PDF

ISO 37500:2014 Guidance on outsourcing

ISO/DIS 59010:2023 Circular Economy. Guidance on the transition of business models and value networks

UNI 10951:2001 Sistemi informativi per la gestione della manutenzione dei patrimoni immobiliari. Linee guida

Chapter 5
Assessing Circular Practices: New Reporting Requirements for the Building Stakeholders

5.1 EU Taxonomy, CSRD and ESRS: New Obligations for Sustainability Reporting

Currently, construction and renovation projects receive solicitations with respect to sustainability issues from two key large areas: (i) the consolidated area of environmental sustainability in the strict sense (including building environmental certifications, product declarations and labels, etc.) and (ii) the new emerging financial area. This second area recently appears to be particularly prolific in regulations and directives and shows an extremely dynamic and evolving regulatory framework which demonstrates the European interest in, and therefore the extreme importance of, the adoption of circular economy and circularity approaches in different sectors, including building design and management which has long been a priority field at European level (Eurostat 2016). In this regard, the main recent regulatory references introduced on a European scale of particular relevance for the building and Real Estate sector are:

– Sustainable Finance Disclosure Regulation (SFDR),[1] which from 2022 has introduced ESG disclosure requirements on sustainability risks and impacts of investment policies and products for financial market participants and advisors. In particular, the key objective of the SFDR is to enable investors to make more informed

[1] See the "Corrigendum to Commission Delegated Regulation (EU) 2022/1288 of 6 April 2022 supplementing Regulation (EU) 2019/2088 of the European Parliament and of the Council with regard to regulatory technical standards specifying the details of the content and presentation of the information in relation to the principle of 'do no significant harm', specifying the content, methodologies and presentation of information in relation to sustainability indicators and adverse sustainability impacts, and the content and presentation of the information in relation to the promotion of environmental or social characteristics and sustainable investment objectives in pre-contractual documents, on websites and in periodic reports (Official Journal of the European Union L 196 of 25 July 2022)", available at https://eur-lex.europa.eu/legal-content/EN/TXT/?uri=CELEX:32022R1288R(01).

N. Atta, *Green Approaches in Building Design and Management Practices*,
Digital Innovations in Architecture, Engineering and Construction,
https://doi.org/10.1007/978-3-031-46760-8_5

investment decisions by requiring financial operators to report comparable ESG data (EY 2023);

- EU Taxonomy, which includes Circular Economy among the 6 key objectives for a sustainable development ("Transition to a Circular Economy" objective). According to the Taxonomy Regulation,[2] the economic activities identified by the EU Taxonomy must contribute substantially to, at least, one of the 6 objectives while also complying with the DNSH—Do No Significant Harm concerning the remaining five objectives (i.e.: climate change mitigation; climate change adaptation; sustainable use and protection of water and marine resources; pollution prevention and control; protection and restoration of biodiversity and ecosystems). The specifications with which the "Transition to a Circular Economy" objective is defined—introduced on 27 June 2023[3]—include specific requirements in terms of Circular Economy for building construction (Table 5.1), renovation (Table 5.2) and demolition (Table 5.3) projects. Furthermore, construction and renovation projects that meet one of the remaining 5 key objectives must in any case demonstrate that they have no negative impact on the environment (DNSH—Do No Significant Harm) and, with respect to the issue of circularity, it is required that "*at least 70% (by weight) of the non-hazardous construction and demolition waste […] generated on the construction site is prepared for reuse, recycling and other material recovery […] in accordance with the waste hierarchy and the EU Construction and Demolition Waste Management Protocol. Operators limit waste generation in processes related to construction and demolition, in accordance with the EU Construction and Demolition Waste Management Protocol and taking into account best available techniques and using selective demolition to enable removal and safe handling of hazardous substances and facilitate reuse and high-quality recycling by selective removal of materials, using available sorting systems for construction and demolition waste. Building designs and construction techniques support circularity and in particular demonstrate, with reference to ISO 20887 or other standards for assessing the disassembly or adaptability of buildings, how they are designed to be more resource efficient, adaptable, flexible and dismantleable to enable reuse and recycling*" (EU 2023a).

---

[2] See Regulation (EU) 2020/852 of The European Parliament and of the Council of 18 June 2020 on the establishment of a framework to facilitate sustainable investment, and amending Regulation (EU) 2019/2088, available at: https://eur-lex.europa.eu/legal-content/EN/TXT/?uri=CELEX:32020R0852.

[3] See Annex II of "Commission Delegated Regulation (EU) […] of 27.6.2023 supplementing Regulation (EU) 2020/852 of the European Parliament and of the Council by establishing the technical screening criteria for determining the conditions under which an economic activity qualifies as contributing substantially to the sustainable use and protection of water and marine resources, to the transition to a circular economy, to pollution prevention and control, or to the protection and restoration of biodiversity and ecosystems and for determining whether that economic activity causes no significant harm to any of the other environmental objectives and amending Delegated Regulation (EU) 2021/2178 as regards specific public disclosures for those economic activities", available at https://finance.ec.europa.eu/system/files/2023-06/taxonomy-regulation-delegated-act-2022-environmental-annex-2_en_0.pdf.

- Corporate Sustainability Reporting Directive (CSRD). Entered into force on 5 January 2023, the CSRD—Directive (EU) 2022/2464—requires companies[4] to report sustainability and circularity data and strategies according to the criteria and indicators defined by the "ESRS—European Sustainability Reporting Standards" developed by EFRAG (European Financial Reporting Advisory Group) with the support of GRI (Global Reporting Initiative). The CSRD is also aligned with the EU Taxonomy requirements. In particular, the companies that fall under the CSRD scope have to report in their annual sustainability reports to what extent their activities (i) are addressed by the EU Taxonomy (Taxonomy-eligibility) and (ii) are in compliance with the requirements and criteria set in the Taxonomy delegated acts (Taxonomy-alignment) (EU 2023b).

Focusing on the CSRD, it is important to highlight its general purposes of (i) improving the quality and comparability of sustainability information, (ii) increasing transparency and accountability and (iii) aligning sustainability reporting with the EU sustainable finance agenda. To this end, the CSRD introduces a novel common standardised reporting framework that is mandatory at the European level (EU 2023a). This reporting framework is based on the recently developed European Sustainability Reporting Standards (ESRS). By requiring the use of ESRS common standards, the CSRD aims to ensure that companies across the EU report comparable and reliable sustainability information. To date, a first set of "Draft ESRS" has been published by EFRAG. In particular, the standards are articulated into:

- Draft ESRS 1 General requirements;
- Draft ESRS 2 General disclosures;
- Draft ESRS E1 Climate change;
- Draft ESRS E2 Pollution;
- Draft ESRS E3 Water and marine resources;
- Draft ESRS E4 Biodiversity and ecosystems;
- Draft ESRS E5 Resources and circular economy;
- Draft ESRS S1 Own workforce;
- Draft ESRS S2 Workers in the value chain;
- Draft ESRS S3 Affected communities;
- Draft ESRS S4 Customers and end-users;
- Draft ESRS G1 Business conduct.

---

[4] The CSRD applies to (EY 2022): (i)Large companies. Companies that meet two of the following conditions: (a) a net turn-over of €40 million; (b) a balance sheet total of €20 million; (c) 250 on average over the financial year. Large companies with substantial business activity in the EU market with a turnover of above €150 million in the EU and at least one subsidiary (large or listed) or branch (with a net turnover of more than €40 million) in the EU; (ii) Small and medium enterprises (SMEs). Small companies with securities listed on regulated markets.

In particular, the CSRD will start to be applied between 2024 and 2028 as follows (EY 2022): from 1 January 2024 for large public-interest companies with over 500 employees, with reports due in 2025 on 2024 data; from 1 January 2025 for large companies with more than 250 employees and/or €40 million of turnover and/or a balance sheet total of €20 million in total assets, with reports due in 2026 on 2025 data; from 1 January 2026 for listed SMEs, with reports due in 2027 on 2026 data. SMEs also have an opt-out option until 2028 to report; from 1 January 2028 for third-country companies, with reports due in 2029 on 2028 data.

Table 5.1 EU Taxonomy: construction of new buildings. Sustantial contribution to the objective "Transition to a Circular Economy". *Source* EU (2023c)

Construction of new buildings
Sustantial contribution to "Transition to a Circular Economy"

1. All generated construction and demolition waste is treated in accordance with Union waste legislation and with the full checklist of the EU Construction and Demolition Waste Management Protocol, in particular by setting sorting systems and pre-demolition audits. The preparing for re-use or recycling of the non-hazardous construction and demolition waste generated on the construction site is at least 90% (by mass in kilogrammes), excluding backfilling. This excludes naturally occurring material referred to in category 17 05 04 in the European List of Waste established by Decision 2000/532/EC. The operator of the activity demonstrates compliance with the 90% threshold by reporting on the Level(s) indicator 2.2 "Construction and demolition waste and materials" using the Level 2 reporting format for different waste streams.
2. The life-cycle Global Warming Potential (GWP) of the building resulting from the construction has been calculated for each stage in the life cycle and is disclosed to investors and clients on demand.
3. Construction designs and techniques support circularity via the incorporation of concepts for design for adaptability and deconstruction as outlined in Level(s) indicators 2.3 "Design for adaptability and renovation" and 2.4 "Design for deconstruction" respectively. Compliance with this requirement is demonstrated by reporting on the Level(s) indicators 2.3 and 2.4 at Level 2.
4. The use of primary raw material in the construction of the building is minimised through the use of secondary raw materials. The operator of the activity ensures that the three heaviest material categories used to construct the building, measured by mass in kilogrammes, comply with the following maximum total amounts of primary raw material used:
   (a) for the combined total of concrete, natural or agglomerated stone, a maximum of 70% of the material come from primary raw material;
   (b) for the combined total of brick, tile, ceramic, a maximum of 70% of the material come from primary raw material;
   (c) for bio-based materials, a maximum of 80% of the total material come from primary raw material;
   (d) for the combined total of glass, mineral insulation, a maximum of 70% of the total material come from primary raw material;
   (e) for non-biobased plastic, a maximum of 50% of the total material come from primary raw material;
   (f) for metals, a maximum of 30% of the total material come from primary raw material;
   (g) for gypsum, a maximum of 65% of the material come from primary raw material.

The thresholds are calculated by subtracting the secondary raw material from the total amount of each material category used in the works measured by mass in kilogrammes. Where the information on the recycled content of a construction product is not available, it is to be counted as comprising 100% primary raw material. In order to respect the Waste Hierarchy and thereby favour re-use over recycling, re-used construction products, including those containing non-waste materials reprocessed on site, are to be counted as comprising zero primary raw material. Compliance with this criterion is demonstrated by reporting in accordance with the Level(s) indicator 2.1 "Bill of Quantities, materials and lifespans".

5. The operator of the activity uses electronic tools to describe the characteristics of the building as built, including the materials and components used, for the purpose of future maintenance, recovery, and reuse, for example using EN ISO 22057:2022 to provide Environmental Product Declarations. The information is stored in a digital format and is made available to investors and clients on demand. In addition, the operator ensures the long-term preservation of this information beyond the useful life of the building by using the information managing systems provided by national tools, such as cadastre or public register.

Source EU (2023c). Available at: https://finance.ec.europa.eu/system/files/2023-06/taxonomy-regulation-delegated-act-2022-environmental-annex-2_en_0.pdf

Table 5.2 EU Taxonomy: renovation of existing buildings. Sustantial contribution to the objective "Transition to a Circular Economy". *Source* EU (2023c)

Renovation of existing buildings
Sustantial contribution to "Transition to a Circular Economy"
1. All generated construction and demolition waste is treated in accordance with Union waste legislation and the full checklist of the EU Construction and Demolition Waste Management Protocol, in particular by setting sorting systems and pre-demolition audits. The preparing for re-use or recycling of the non-hazardous construction and demolition waste generated on the construction site is at least 70% (by mass in kilogrammes), excluding backfilling. This excludes naturally occurring material referred to in category 17 05 04 in the European List of Waste established by Commission Decision 2000/532/EC. The operator of the activity demonstrates compliance with the 70% threshold by reporting on the Level(s) indicator 2.2 "Construction and demolition waste and materials" using the Level 2 reporting format for different waste streams. 2. The life cycle Global Warming Potential (GWP) of the building renovation works has been calculated for each stage in the life cycle, from the point of renovation, and is disclosed to investors and clients on demand. 3. Construction designs and techniques support circularity via the incorporation of concepts for design for adaptability and deconstruction as outlined in Level(s) indicators 2.3 "Design for adaptability and renovation" and 2.4 "Design for deconstruction" respectively. The operator of the activity demonstrates compliance with this requirement by reporting on the Level(s) indicators 2.3 and 2.4 at Level 2. 4. At least 50% of the original building is retained. This is to be calculated based on the gross external floor area retained from the original building using the applicable national or regional measurement methodology, alternatively using the definition of 'IPMS 1' contained in the International Property Measurement Standards. 5. The use of primary raw material in the renovation of the building is minimized through the use of secondary raw materials. The operator of the activity ensures that the three heaviest material categories that have been newly added to the building in the renovation of the building, measured by mass in kilogrammes, comply with the following thresholds regarding the maximum amount of primary raw material used: (a) for the combined total of concrete, natural or agglomerated stone, a maximum of 85% of the material come from primary raw material; (b) for the combined total of brick, tile, ceramic, a maximum of 85% of the material come from primary raw material; (c) for bio-based materials, a maximum of 90% of the material come from primary raw material; (d) for the combined total of glass, mineral insulation, a maximum of 85% of the material come from primary raw material; (e) for non-biobased plastic, a maximum of 75% of the material come from primary raw material; (f) for metals, a maximum of 65% of the material come from primary raw material; (g) for gypsum, a maximum of 83% of the material come from primary raw material. The thresholds are calculated by subtracting the secondary raw material from the total amount of each material category used in the works measured by mass in kilogrammes. Where the information on the recycled content of the construction product is not available, it is to be counted as comprising 100% primary raw material. In order to respect the Waste Hierarchy and thereby favour re-use over recycling, re-used construction products, including those containing non-waste materials reprocessed on site, are to be counted as comprising zero primary raw material. Compliance with this criterion is demonstrated by reporting in accordance with the Level(s) indicator 2.1 "Bill of Quantities, materials and lifespans". The operator of the activity uses electronic tools to describe the characteristics of the building as built, including the materials and components used, for the purpose of future maintenance, recovery, and reuse, for example using EN ISO22057:2022 to provide Environmental Product Declarations111. The information is stored in a digital format and is made available to investors and clients on demand. In addition, the operator of the activity ensures the long-term preservation of this information beyond the useful life of the building by using the information managing systems provided by national tools, such as cadastre or public register.

Source EU (2023c). Available at: https://finance.ec.europa.eu/system/files/2023-06/taxonomy-regulation-delegated-act-2022-environmental-annex-2_en_0.pdf

The Draft ESRS E5 concerns circular economy and circularity and it defines different topical Disclosure Requirements related to "*(a) resource inflows including the circularity of material resource inflows, considering renewable and non-renewable resources; and (b) resource outflows including information on products and materials; and (c) waste*" (EFRAG 2022a) introduced in Table 5.4.

As it is possibile to observe reading the Discolsure Requirements of ESRS E5 (Table 5.4), the information and data to disclose within sustainability reports address the reporting of sustainability impacts from two points of view, according to the

Table 5.3 EU Taxonomy: demolition of buildings. Sustantial contribution to the objective "Transition to a Circular Economy". *Source* EU (2023c)

Demolition and wrecking of buildings and other structures
Sustantial contribution to "Transition to a Circular Economy"
1. Prior to the start of the demolition or wrecking activity, at least the following aspects from the Level 1 design concept checklist of the Level(s) indicator 2.2 "Construction and demolition waste and materials" checklist are discussed and agreed upon with the client: (a) definition of key performance indicators and target ambition level; (b) identification of project-specific constraints that may compromise the target ambition level (such as time, labour and space) and how to minimise these constraints; (c) details of the pre-demolition auditing procedure; (d) an outline waste management plan that prioritises selective deconstruction, decontamination and source separation of waste streams. Where these actions are not prioritised, an explanation is provided to justify why selective deconstruction, decontamination or source separation of waste streams are not technologically feasible in the project. Cost or financial considerations are not an acceptable reason to avoid complying with this requirement. 2. The operator of the activity conducts a pre-demolition audit in line with the EU Construction and Demolition Waste Management Protocol. 3. All demolition waste generated during the demolition or wrecking activity is treated in accordance with Union waste legislation and the full checklist of the EU Construction and Demolition Waste Management Protocol. 4. The preparing for re-use or recycling of the non-hazardous construction and demolition waste generated on the construction site is at least 90% (by mass in kilogrammes), excluding backfilling. This excludes naturally occurring material referred to in category 17 05 04 in the European List of Waste established by Commission Decision 2000/532/EC. The operator of the activity demonstrates compliance with the 90% threshold by reporting on the Level(s) indicator 2.2 "Construction and demolition waste and materials" using the Level 3 reporting format for different waste streams. Alternatively, at least 95% of the mineral fraction and 70% of the non-mineral fraction of the non-hazardous demolition waste is separately collected and prepared for reuse or recycled.

Source EU (2023c). Available at: https://finance.ec.europa.eu/system/files/2023-06/taxonomy-regulation-delegated-act-2022-environmental-annex-2_en_0.pdf

Table 5.4 EU sustainability reporting standards: ESRS E5 Resource use and circular economy by EFRAG. *Source* EFRAG (2022a)

ESRS E5 Resource use and circular economy

Discolsure requirement	***Description (extract**—source: [Draft] ESRS E5 Resource use and circular economy.* EFRAG 2022a)
Disclosure requirement related to [draft] ESRS 2 IRO-1—**Description of the processes to identify and assess material resource use and circular economy-related impacts, risks and opportunities**	The undertaking shall describe the process to identify material impacts, risks and opportunities related to resource use and circular economy and shall provide information on: (a) the methodologies, assumptions and tools used to screen its assets and activities in order to identify its actual and potential risks in its own operations and value chain; (b) the interconnection between risks and opportunities arising from impacts and dependencies; (c) the process for conducting consultations, in particular with affected communities.
Disclosure requirement E5-1—**Policies related to resource use and circular economy**	The undertaking shall disclose its policies implemented to manage its material impacts, risks and opportunities related to resource use and circular economy. The objective of this Disclosure Requirement is to enable an understanding of the extent to which the undertaking has policies that address the identification, assessment, management and/or remediation of its material impacts, risks and opportunities related to resource use and circular economy. The summarised description of the policy shall contain the information required in [draft] ESRS 2 DC-P Policies adopted to manage material sustainability matters. In the summary, the undertaking shall indicate whether and how its policies address the following matters where material: (a) transitioning away from extraction of virgin non-renewable resources; (b) securing and contributing to the regenerative production of renewable resources and the regeneration of ecosystems they are part of. Policies shall address material impacts, risks and opportunities in its own operations and along the upstream and downstream value chain.

(continued)

Table 5.4 (continued)

ESRS E5 Resource use and circular economy	
Discolsure requirement	***Description (extract**—source: [Draft] ESRS E5 Resource use and circular economy.* EFRAG 2022a)
Disclosure requirement E5-2—**Actions and resources related to resource use and circular economy**	The undertaking shall disclose its resource use and circular economy actions and the resources allocated to their implementation. The objective of this Disclosure Requirement is to enable an understanding of the key actions taken and planned to achieve the resource use and circular economy-related policy objectives. The description of the resource use and circular economy-related action and resources allocated shall follow the principles defined in [draft] ESRS 2 DC-A Actions and resources in relation to material sustainability matters. In addition to [draft] ESRS 2 DC-A, the undertaking shall specify whether and how an action and resources cover: (a) any of the layers of the waste hierarchy as defined in Appendix A; (b) more detailed circular economy strategy throughout the value chain of the product: Refuse, Rethink, Reduce, Reuse, Repair, Refurbish, Remanufacture and Repurpose), Recycle. The disclosure shall also include a description of the actions, including circularity measures, taken to prevent waste generation in the undertaking's upstream and downstream value chain and to manage material impacts arising from waste generated.
Disclosure requirement E5-3—**Targets related to resource use and circular economy**	The undertaking shall disclose the resource use and circular economy-related targets it has adopted. The objective of this Disclosure Requirement is to enable an understanding of the targets the undertaking has adopted to support its resource use and circular economy policy and to address its material impacts, risks and opportunities. The description of the targets shall contain the information requirements defined in [draft] ESRS 2 DC-T Tracking effectiveness of policies and actions through targets. The disclosure shall indicate whether and how its targets relate to inflows and outflows, including waste and products and materials, (including in production, use phase and at end of functional life) and, more specifically to: (a) the increase of circular design (including for instance product design); (b) the increase of circular material use rate; (c) the minimisation of virgin non-renewable raw material with possibly targets for virgin non-renewable raw material and targets for virgin renewable raw material; (d) the reversal of the depletion of the stock of renewable resources; (e) the waste management, including preparation for proper treatment; and (f) other targets. The undertaking shall specify to which layer of the waste hierarchy the target relates. In addition to [draft] ESRS 2 DC-T, the undertaking shall specify whether (local) ecological thresholds and entity-specific allocations were taken into consideration when setting targets. If so, the undertaking should specify: (a) the ecological thresholds identified, and the methodology used to identify such thresholds; (b) whether or not the thresholds are entity-specific and if so, how they were determined; and (c) how responsibility for respecting identified ecological thresholds is allocated in the undertaking. The undertaking shall specify as part of the contextual information, whether the targets it has adopted and presented are mandatory (based on legislation) or voluntary and if and how such legal requirements were taken into account when considering ecological thresholds.

(continued)

Table 5.4 (continued)

ESRS E5 Resource use and circular economy

Discolsure requirement	***Description (extract****—source: [Draft] ESRS E5 Resource use and circular economy.* EFRAG 2022a)
Disclosure requirement E5-4—**Resource inflows**	The undertaking shall disclose information on its material resource inflows. The objective of this Disclosure Requirement is to enable an understanding of the resource use in the course of the undertaking's own operations and value chain. The disclosure shall include a description of its material inflows: products (including packaging) and materials, and property, plant and equipment used in the undertaking's own operations and along the value chain. For undertakings for which inflows are material and those active in one of "key products value chain", as defined in the EU Circular Economy action plan, the undertaking shall include, in tonnes or kilo, at the reporting period: (a) the overall total weight of products and materials used during the reporting period; (b) the weight in both absolute value and percentage of renewable input materials from regenerative sources used to manufacture the undertaking's products and services (including packaging); and (c) the weight in both absolute value and percentage, of reused or recycled products and materials (non-virgin) used to manufacture the undertaking's products and services (including packaging). The undertaking shall provide information on the methodologies used to calculate the data. It shall specify whether the data is sourced from direct measurement or estimations, and disclose the key assumptions used.
Disclosure requirement E5-5—**Resource outflows**	The undertaking shall disclose information on its material resource outflows, including waste. The objective of this Disclosure Requirements is to provide an understanding of: (a) how the undertaking contributes to circular economy by i) designing products and materials in line with circular principles and ii) the extent to which products, materials and waste processing are recirculated in practice after first use; and (b) the undertaking's waste management strategy and the extent to which the undertaking knows how its waste is managed in its own activities.

(continued)

Table 5.4 (continued)

ESRS E5 Resource use and circular economy	
Discolsure requirement	***Description (extract**—source: [Draft] ESRS E5 Resource use and circular economy.* EFRAG 2022a)
	Products and materials The undertaking shall provide a description of the key products and materials that come out of the undertaking's production process and that are designed along circular principles, including durability, reusability, repairability, disassembly, remanufacturing, refurbishment, recycling or other optimisation of the use of the resource. The undertaking for which outflows are material and those active in one of "key products value chain" as defined in the EU Circular Economy action plan, shall provide information at the reporting period on: (a) the total weight (tonnes) and percentage of materials that come out of the undertaking's products and services production process (including packaging) that have been designed along circular principles: i. durability; ii. reusability; iii. repairability; iv. disassembly; v. remanufacturing or refurbishment; vi. recycling; vii. recirculation by the biological cycle; viii. other potential optimisation of product and material use; and (b) the weight and percentage of products and materials that come out of the undertaking including packaging that are designed to enhance/enable circular economy for customers further down the value chain.
	Waste The undertaking shall disclose the following information on its total amount of waste on its own operations at the reporting period, in tonnes or kilogrammes: (a) the total amount of waste generated; (b) for each type of hazardous and non-hazardous waste, the amount by weight diverted from disposal by recovery operation type and the total amount summing all three types. The recovery operation types to be reported on are: i. preparation for reuse; ii. recycling; and iii. other recovery operations; (c) for each type of hazardous and non-hazardous waste, the amount by weight directed to disposal by waste treatment type and the total amount summing all three types. The waste treatment types to be disclosed are: i. incineration; ii. landfilling; and iii. other disposal operations; (d) the total amount and percentage of non-recycled waste. When disclosing the composition of the waste, the undertaking shall specify: (i) the waste streams, relevant to its sector or activities (e.g. tailings for the undertaking in the mining sector, electronic waste for the undertaking in the consumer electronics sector, or food waste for the undertaking in the agriculture or in the hospitality sector); and (ii) the materials that are present in the waste (e.g. biomass, metals, non-metallic minerals, plastics, textiles). The undertaking shall also disclose the total amount of hazardous waste and radioactive waste generated by the undertaking, where radioactive waste is defined in Article 3 of Council Directive 2011/70/Euratom. The undertaking shall provide contextual information on the methodologies used to calculate the data and in particular the criteria and assumptions used to determine and classify products designed along circular principles. It shall specify whether the data is sourced from direct measurement or estimations; and disclose the key assumptions used.

(continued)

Table 5.4 (continued)

ESRS E5 Resource use and circular economy	
Discolsure requirement	***Description (extract**—source: [Draft] ESRS E5 Resource use and circular economy.* EFRAG 2022a)
Disclosure requirement E5-6—**Potential financial effects from resource use and circular economy-related impacts, risks and opportunities**	The undertaking shall disclose its potential financial effects of material risks and opportunities arising from resource use and circular economy-related impacts. The objective of this Disclosure Requirement is to provide an understanding of: (i) potential financial effects due to material risks arising from resource use and circular economy-related impacts and dependencies and how these risks may have a material influence (or are likely to have a material influence) on the undertaking's cash flows, performance, position, development, cost of capital or access to finance over the short-, medium- and long-term time horizons; and (ii) potential financial effects due to material opportunities arising from resource use and circular economy-related material impacts and how the undertaking may financially benefit from material resource use and circular economy-related opportunities. The disclosure shall include: (a) a quantification of the potential financial effects in monetary terms, or where impracticable qualitative information. For financial effects arising from material opportunities, a quantification is not required if it would result in disclosure that does not meet the qualitative characteristics of information (see [draft] ESRS 1 Appendix C Qualitative characteristics of information); (b) a description of the considered effects, the impacts to which they relate and the time horizons in which they are likely to materialise; (c) the critical assumptions used in the estimate, as well as the sources and level of uncertainty attached to those assumptions. In the context of this Disclosure Requirement, potential financial effects include financial effects that do not meet the recognition criteria for inclusion in the financial statement line items and notes to the financial statements.

so-called "double materiality perspective",[5] namely: (i) on one side, companies are required to report their impacts on people and the environment and (ii) on the other hand, companies are required to assess and report how social and environmental sustainability issues generate financial effects, impacts, risks and opportunities for the company itself and its business (EFRAG 2022b). Hence, the "impact of" and "impacts on" the company, therefore, are each considered a materiality perspective.

Morevoer, it is important to highlight how the CSRD includes in the scope of the sustainability report, thus of the ESG assessment, not only the companies themselves but also their whole supply chains. Hence, the reporting activities must also include the assessment of sustainability along the entire vaule chain of the companies.

These novelties may imply for the different operators of the building sector (project developers, building design companies, construction companies, building product manufacturers, etc.) to:

- integrate environmental materiality into their reporting. Companies now need to counterweight their financial performance with their environmental performance, adopting the environmental metrics of the ESRS (including circularity metrics—see Table 5.4);
- define a long-term ESG strategy and include into sustainability management practices the periodic assessment of environmental sustainability risks, effects and opportunities with impacts both internally (company itself) and externally (external system on which the company can have an impact);
- establish a network with all the players of the value chain in order to increase the transparency and the understanding of sustainability risks and opportunities. Also in relation to "green finance", this aspect is crucial especially in construction projects where companies with a reporting obligation often depend on their suppliers and partners (supply chain). Therefore, establishing a partership with the different value chain stakeholders—facilitating ESG data collection and sharing—become fundamental;
- acquire methods and tools useful for increasing the availability of ESG data within the company and along the company supply network, including digital information systems and structured procedures for collecting, sharing, analyzing and archiving data useful for calculating or estimating ESRS (including those in Table 5.4 concerning the ESRS E5 "Resource use and circular economy"), thus laying the foundation for the development of sustainability reports according to the new rules set by the CSRD.

[5] "*Double materiality is a concept which provides criteria for determination of whether a sustainability topic or information has to be included in the undertaking's sustainability report. Double materiality is the union (in mathematical terms, i.e. union of two sets, not intersection) of impact materiality and financial materiality. A sustainability topic or information meets therefore the criteria of double materiality if it is material from the impact perspective or from the financial perspective or from both of these two perspectives*" (EFRAG 2022b).

5.2 Circularity Assessment: Review of Voluntary Tools and Support Guidelines

The present paragraph aims to introduce and review virtuous voluntary guidelines and support tools for circularity assessment, applicable at different levels (product, process, organization) in the building context for evaluating the level of circularity of the practices carried out during the design and management phases of the Building Process. In particular, the tools introduced in the following sub-paragraphs are: (i) ISO/DIS 59020 standard "Circular economy. Measuring and assessing circularity" by the International Organization for Standardization (ISO); (ii) Circulytics by the Ellen MacArthur Foundation (EMF) and (iii) the Circular Transition Indicators (CTI) by the World Business Council for Sustainable Development (WBCSD).

5.2.1 *The ISO/DIS 59020 International Standard: Towards a Common Guideline to Measure and Assess Circularity*

The ISO/DIS 59020 standard "Circular economy. Measuring and assessing circularity" is the most recent among the standards composing the ISO 59000 series, developed by the International Organization for Standardization (ISO). The ISO/DIS 59020 aims to provide a structured approach to measure and assess circularity performance and sustainability impacts based on standard indicators and related measuring methods (ISO/DIS 59020:2023). In particular, the standard: "*provides guidance to organizations for measuring and assessing a selected system to determine its circularity performance at a specific moment in time. Measurement and assessment are performed by collection and calculation of data with the help of circularity indicators and appropriate complementary methods such as life cycle assessment. [...] A framework for organizations to measure and assess circularity, enabling those organizations to contribute to sustainable development. The framework is applicable to multiple levels of an economic system, ranging from regional, interorganizational, organizational to the product level*" (ISO/DIS 59020:2023).

The framework for measuring and assessing circularity proposed by the standard is articulated into three key interrelated procedural steps (Fig. 5.1). The outcome of the application of the framework, thus of the measurement and assessment process, is specific to a defined moment in time (or timeframe) for which the measurement and assessment is conducted. Hence, to monitor the progress over time, the framework should be re-applied regularly (periodic measurement of key indicators). The measurement and assessment process can be carried out using a digital information system (or information platform), relying on an ICT infrastructure to streamline the whole process and to enhance the accuracy of the related outcomes.

The first step of the procedure—"1. Boundary setting"—is represented by the definition of the application context, by identifying the scope and setting the boundaries of the "system" to be measured and assessed. To this end, the standard proposes

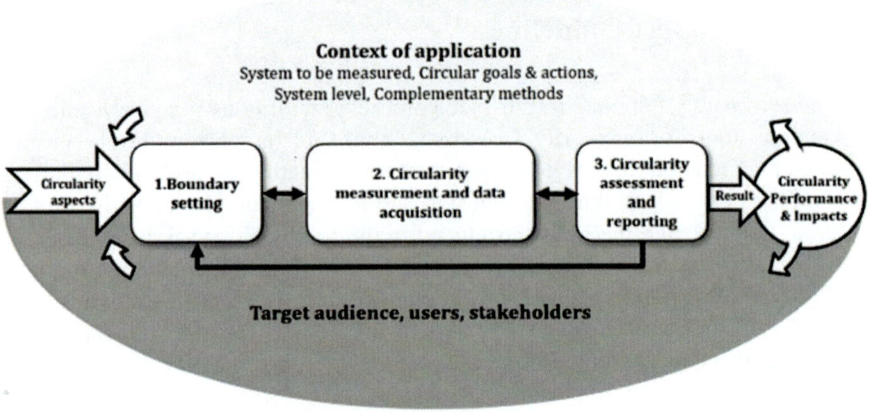

Fig. 5.1 Framework for measuring and assessing circularity by ISO/DIS 59020. *Source* ISO/DIS 59020:2023

a list of activities to perform and a set of questions to reply by carrying out the activities themselves, as shown in Table 5.5. In particular, this first step involves: (i) the definition of the "system in focus" (ISO/DIS 59020:2023), meant as the subject of the circularity measurement and assessment; (ii) the identification of circularity aspects to measure, e.g. dis/assemblability, maintainability, repairability, reusability, recyclability, etc.; (iii) the definition of data quality requirements; and (iv) the pre-selection of methods and tools for the measurement and assessment of the social, environmental, and economic impacts (ISO/DIS 59020:2023). Moreover, in this first step, it is also important to set the temporal boundary (namely the moment or timeframe of reference for the analysis) and the level of observation, i.e. regional, inter-organizational or product level (ISO/DIS 59020:2023).

The second step focuses on the definition of the indicators and the related required data collection. In particular, the standard defines circularity indicators as "*quantitative or qualitative measure of a circularity aspect*" (ISO/DIS 59020:2023). Moreover, it specifies that indicators "*can be used at any life cycle phase of a solution. This includes for example, during simulation to test circular designs or during production, use, remanufacturing, repurposing and recovering*" (ISO/DIS 59020:2023).

The process of measuring circularity and data acquisition proposed by ISO/DIS 59020 consists of different sub-steps (namely: choose circularity indicators, identify information to be measured, acquire data, calculate and/or aggregate, review the selection if data cannot be acquired) with possible iterations as shown in Fig. 5.2. Hence, this second step is strongly related with the first and third procedural steps towards the achievement of continuous adjusting and improvement (Fig. 5.3).

To measure and assess circularity performance of a system in focus (product, solution, organization, value network, etc.), the ISO/DIS 59020 standard provides a

Table 5.5 Framework for measuring and assessing circularity by ISO/DIS 59020. Step 1: Boundary setting. *Source* ISO/DIS 59020:2023

Setting the boundaries for the circularity assessment Source: ISO/DIS 59020:2023
Key questions to reply
What is the purpose and intended use of the result of the circularity measurement and assessment? *Which system level (regional, inter-organizational, organization, product) is applicable?* *What circular goals and targets and resource flows should be measured and assessed?* *What circularity aspects such as repairability, reusability and durability are of specific interest and importance for taking action and achieving circular goals?* *What requirements and goals for specific social, environmental, and economic issues or Sustainable Development Goal (SDG) targets should be considered for possible impacts?* *What value network actors or stakeholders are part of the system in focus and its boundaries to be included and how to share information? What is the purpose of the report (e.g. comparative assertions, benchmarking, public disclosure, information for stakeholders, internal communication)?*
Determining goal and scope
• Defining circular objectives and goals to be measured (e.g. increase repair of product A with X %); • reduce non-renewable content of the organization with X %, increase recyclability of plastic in a city with X %), including the circularity aspects (e.g. repairability, recyclability, reusability, durability, recovered losses, etc.); • defining the system in focus, i.e. which system level, functional units, locations, parts of a value chain or value network, regions etc. are to be measured and assessed; • defining data quality requirements (for example X % of acquired data needs to be primary); • defining pre-selection of complementary methods for social, environmental and economic impact measurement and assessment (e.g. ISO 14044 for LCA, ISO 26000 for social responsibility); • defining the stakeholders involved, i.e., the users or practitioners carrying out or using the measurement and assessment, the target audience (to whom the results of the assessment will be communicated, internal and/or external) and other stakeholders (related to acquiring the data and assessing the impacts and organizations outside the system in focus that can be impacted e.g. companies within the value chain); • defining whether the results are to be used in comparative circularity assertions (i.e. claims that a product, solution, organization, process, value network etc. is more circular than a competing one) that are intended to be disclosed to the public; • defining the boundaries of the system in focus with the wider economic system and social and environmental systems.
Defining the boundaries of the system
Take into consideration: • all resource flows associated with the system in focus (e.g. specify types and amounts of primary and secondary resources); • criteria and rules for boundary setting (such as how to distinguish and account for resource flows and stocks, or how to assign quantitative or qualitative values to resources); • the methodology to be used (such as normalization to a specific reference–year, product, etc.–comparison to similar systems, sensitivity analyses); • data quality considerations (such as time specific data, which data are primary data and sources of secondary data); • consideration of the sources and availability of data such as suppliers and organizations within the value chain or value network; • assumptions such as technology readiness level, market size, resource lifetime, waste management systems pricing, regulatory incentives etc., under technical, economic and regulatory criteria that can influence the circularity performance; • how value(s) can be appraised including economic, social and environmental contributions.

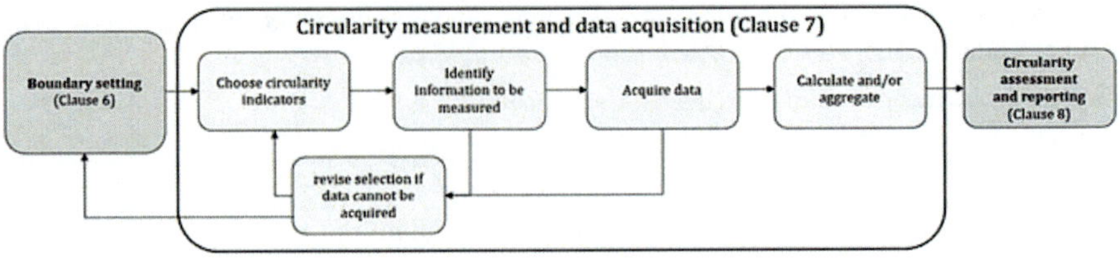

Fig. 5.2 Sub-steps of the circularity measurement and data acquisition process by ISO/DIS 59020. *Source* ISO/DIS 59020:2023

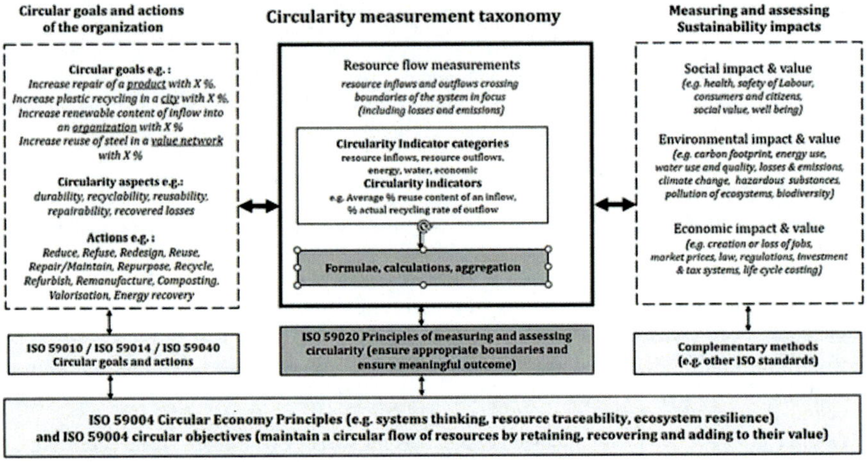

Fig. 5.3 Circularity measurement taxonomy by ISO/DIS 59020 and interactions among steps and activities. *Source* ISO/DIS 59020:2023

taxonomy (Fig. 5.3) of circularity indicator categories and indicators (Table 5.6).[6] From the proposed circularity indicators, the operators can preliminarily identify the data and information to collect, measure and/or calculate.

Lastly, the third step, i.e. Circularity assessment and reporting, concerns the evaluation of the results of the circularity indicators calculation. The aim of this step is to analyse, compare (e.g. using benchmarks and expected values) and interpret the values of the calculated circularity indicators in order to report the overall and detailed circularity performance of the system in focus. The assessment can take into account the degree of matching between the adopted circular practices and the key circular economy objectives and principles, expressed for instance in the standard ISO/DIS 59004 "Circular Economy. Terminology, principles and guidance for implementation". Moreover, the assessment should also include the estimation of sustainability impacts, assessed in terms of social (e.g. new job opportunities deriving from the implementation of remanufacturing strategies, how changes in resource use

[6] See Annex A "Core circularity indicators and data measurement" of ISO/DIS 59020:2023 for the technical specifications and formulas to calculate the circularity indicators proposed by ISO/DIS 59020:2023, shown in Table 5.6.

Table 5.6 Circularity indicator categories and core circularity indicators by ISO/DIS 59020. *Source* ISO/DIS 59020:2023

Indicator category	Circularity indicator	Summary description
Resource inflows	Average percent reused content of an inflow	Fraction of input material resources that is reused content
	Average percent recycled content of an inflow	Fraction of input material resources that is recycled content
	Average percent renewable content of an inflow	Fraction of material resources inflow that is sustainably produced renewable content
Resource outflows	Average lifetime of product or material relative to industry average	Indicator of time that an output resource (e.g. product) will remain in use compared to an industry average for the resource
	Percent actual reused content derived from outflow	Fraction of outflow that is reused
	Actual % recycling rate of outflow	Fraction of outflow that is recycled
	Percent actual recirculation of outflow in the biological cycle	Fraction of outflow content that is recirculated at end of life for safe return to the biosphere and meets the qualifying conditions for recirculation
Energy	Average percent of energy consumed that is renewable energy	Fraction of net consumed energy that qualifies as renewable energy, taking into account both energy inflows and energy outflows
Water	Percent water withdrawal from circular sources	Percent of annual water demand that is derived from circular sources
	Percent water discharged in accordance with quality requirements	Percent (by volume) of total water withdrawn that is discharged in accordance with circularity principles
	Ratio (onsite or internal) water reuse or recirculation	Reuse cycles of onsite water
Economic	Revenue share of circular resources (or products) (RSCR)	Percent of total revenue generated per year by use of circular (and/or) non-circular resources
	Material productivity (MP)	Ratio of revenue generated by total mass of all linear resource inflows
	Resource intensity index (RII)	Quantitative measure of economic growth versus total resource use

can negatively/positively affect local employment rates, etc.) environmental (e.g. avoided emissions due to the reduced amount of waste to landfill, etc.) and economic (e.g. reduction of the costs for accessing to products or services related to sharing or as-a-service models; circular re-strategies as repair, remanufacture and repurposing can affect prices of resources as well as new and used products, etc.) terms (ISO/DIS 59020:2023).

In particular, this step is articulated in the sub-steps shown in Fig. 5.4, namely: review measurement results (including calculation and aggregation of results); assessing value and impact (social, environmental and economic); consult stakeholders, users and target audience on the results; document and report the circularity performance outcome producing a circularity assessment report.

With respect to the operators of the building sector, the ISO/DIS 59020 represents a useful framework of common reference that can guide towards the implementation and assessment of circular practices. Indeed, the standard provides simplified methods and tools for setting circularity goals and assessing the efficacy and efficiency of the implemented sustainable practices. The ISO/DIS 59020 standard, together with the other standards of the ISO 59000 series, represents a fundamental reference for supporting building operators, especially small and medium-size companies, through the gradual implementation and assessment of circular practices.

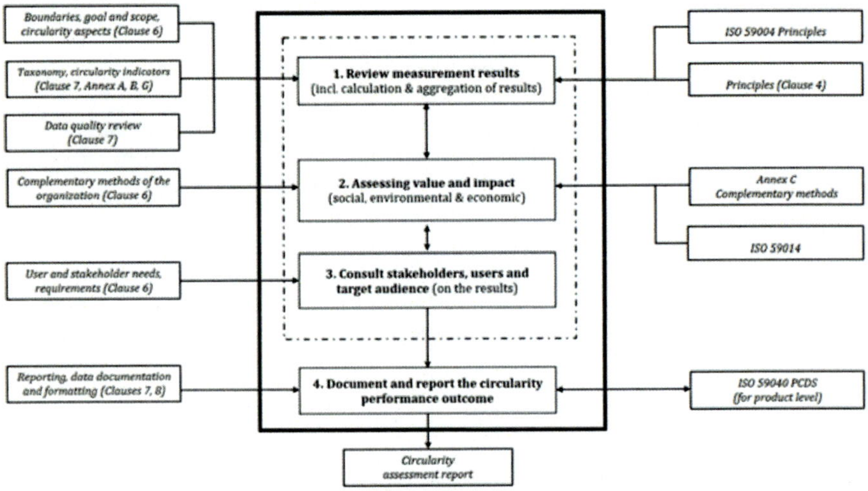

Fig. 5.4 Sub-steps of Circularity assessment and reporting process by ISO/DIS 59020. *Source* ISO/DIS 59020:2023

5.2.2 *Circulytics by Ellen MacArthur Foundation*

Circulytics (launched in 2020) is a tool developed by Ellen MacArthur Foundation to assess circular economy performance of companies' operations. The tool provides Circular Economy indicators articulated into two categories: 1. Enablers and 2. Outcomes.

In particular, the category 1 "Enablers" highlights the key aspects that enable a company to shift towards circular practices, including strategic prioritisation of Circular Economy and development of support systems for circular operations. This category is divided into 5 themes, namely: strategy and planning; innovation; people and skills; operations; external engagement. For each theme, the tool proposes a set of relevant indicators (Table 5.7) expressed in the form of questions with predefined answers to choose from.[7] The tool then includes a weighing system to translate in quantitative terms the outcomes of the set of indicators (company-selected answers).

The second category "2. Outcomes" assesses the actual Circular Economy results of a company according to six themes, namely: products and materials; services; plant, property, and equipment assets; water; energy; finance. For each theme, the tool proposes a set of key indicators (Table 5.8) for the circularity assessment. As for the category 1 "Enablers", also for this second category the indicators are expressed in the form of questions with predefined answers to choose from.[8] Through a weighing system (a weight is attributed to each question and each answer) the indicators are then translated from qualitative to quantitative.

From 2020, the EMF counts more than 2,000 companies signed up to Circulytics to have access to the data-driven tool in order to measure their Circular Economy performance and outline ways to improve current circular strategies. Recently, the EMF—following the launch of the CSRD (Corporate Sustainability Reporting Directive) by EU and the related introduction of the European Sustainability Reporting Standards (ESRS)—decided that from 31 August 2023 will not receive any more submission from companies for the drafting of the sustainability report.[9] Besides, by

---

[7] See the complete EMF Circulytics tool ("Enablers" Indicator List) at https://emf.thirdlight.com/link/1pzbxosbi6hl-ei3tq6/@/#id=2.

[8] See the complete EMF Circulytics tool ("Outcomes" Indicator List) at https://emf.thirdlight.com/link/1pzbxosbi6hl-ei3tq6/@/#id=2.

[9] "*More recently, the adoption of circular economy indicators in mandatory reporting frameworks is a welcome development, such as the forthcoming European Sustainability Reporting Standards (ESRS) which the Foundation has informed and is in good alignment with many Circulytics indicators. […] Given the fast-changing non-financial reporting landscape, many companies will be concentrating effort on regulated disclosure requirements. Now is the time for harmonisation and streamlining of reporting. The Foundation will therefore be stepping away from data collection and individual performance assessments based on Circulytics. This means we will not accept new Circulytics submissions after 31st August 2023. The Circulytics methodology and resources will remain available on our website for reference, and we will explore opportunities to enhance its value as a tool for the preparation of disclosures. As a priority, we encourage organizations to disclose their circular economy performance as outlined in the ESRS, for their company's global scope. To support this, we have published a two-way mapping of the ESRS Drafts and Circulytics indicators*" (EMF 2023).

Table 5.7 Circulytics indicators for the category "Enablers". *Source* EMF (2022)

Circulytics—indicator category 1. Enablers (*Source* EMF 2022)		
Theme	General	Indicator
1. Strategy and planning	Have you placed the circular economy at the heart of your strategy?	1a. How central is circular economy to your CEO's agenda?
		1b. Does your organisational risk management include risks and opportunities related to the transition to a circular economy, and the risks of staying in a linear economy?
		1c. Is your strategy aligned with becoming more circular?
		1d. Do you have measurable circular economy targets?
		1e. Are the following publicly available (e.g. in an annual report)?
		1f. Do you have a circular economy implementation plan?
2. Innovation	Are the conditions in place to support the development of innovative circular products and services? Are you innovating towards new CE products, systems, or services?	2a. To what extent is leadership involved in supporting circular innovation/development projects?
		2b. To what extent are tools and metrics in place to support circular innovation/development projects?
		2c. To what extent do you collaborate on circular innovation/development projects?
		2d. To what extent are different data systems in place to support circular innovation/ development projects?
3. People and skills	Are you supporting your employees? Have you employed people to develop the skills required to transition to a circular business model?	3a. To what extent are your circular economy strategy and implementation plans communicated internally?
		3b. To what extent does your company offer circular economy related training within your company?
		3c. In which functions do you have individuals or project teams with responsibility for circular economy implementation?

(continued)

Table 5.7 (continued)

Circulytics—indicator category 1. Enablers (*Source* EMF 2022)		
Theme	General	Indicator
4. Operations	Have you invested sufficiently in your digital systems and assets to support the change?	4a. To what extent is your company implementing digital systems to support circular products or circular services?
		4b. To what extent is your company implementing plant, property, and equipment assets to support circular products or circular services?
5. External engagement	Are you promoting your circular economy initiatives and influencing those in your business sphere, such as clients or your supply chain?	5a. To what extent do you engage with suppliers to increase sourcing based on circular economy principles?
		5b. To what extent do you engage with customers on advancing circular economy topics?
		5c. To what extent do you engage with policymakers to support the transition to a circular economy?
		5d. To what extent do you engage with external investors and/or financiers of your company on circular economy topics?
		5e. Do you have a membership of or actively engage with circular economy related initiatives?

highlighting key Circular Economy priorities for company operations, Circulytics can still be a reference tool that can be of fundamental importance especially for the small and medium building operators, for which reporting is not yet mandatory, to gradually start adopting circularity principles in their practices and prepare for the future drafting of the sustainability report and related disclosure requirements according to the ESRS as per CSRD. To this end, EMF has made available a comparative table between the Circulytics indicators and the ESRS-5 "Resource use and circular economy" (available at https://ellenmacarthurfoundation.org/circulytics-esrs) (EMF 2023).

5.2.3 *Circular Transition Indicators (CTI) by WBCSD*

The Circular Transition Indicators (CTI)—developed by the World Business Council for Sustainable Development (WBCSD)—provides companies a set of indicators

Table 5.8 Circulytics indicators for the category "Outcomes". *Source* EMF (2022)

Circulytics—indicator category 2. Outcomes (*Source* EMF 2022)		
Theme	General	Indicator
6. Products and materials	Are the materials you procure and the products you design supporting a circular economy?	6a. What % (by mass) of products and materials (in-flows) is: Non-virgin (e.g. reused and recycled products and materials); Sourced from by-products/waste streams (e.g. offcuts of a material that has not previously been in a product); Virgin but renewable and regeneratively produced; Virgin but renewable and sustainably produced (products and materials that are produced sustainably, but not regeneratively)?
		6b. What % (by mass) of waste, is material processing waste or by-products that go to landfill or incineration (and are therefore not recirculated)?
		6c. What % (by mass) of waste is materials processing waste or by-products that go to landfill or incineration (and are therefore not recirculated)?
		6d. What % (by mass) of your physical products are designed along circular economy principles?
		6e. Do your product and material outflows (all products, packaging, material processing waste and by-products) comply with either of the following chemical restriction lists?
		6f. Part 1. What % (by mass) of your products and materials are recirculated in practice in the following ways (only counting the first cycle of recirculation after initial use): Reuse/redistribution, Refurbishment/remanufacture, Recycling, Nutrient recirculation that meets the qualifying conditions (e.g. composting and anaerobic digestion) 6f. Part 2. For all products and materials that are recirculated through reuse, how many average uses do your products have before reaching the end of functional life?
7. Services	Are the services you provide supporting a circular economy?	7a. Part 1. What % of your service revenue is from circular services? 7a. Part 2. Select the circular economy principle(s) that the services you highlighted in Part 1 have a positive impact on, and describe the impact: Eliminate waste and pollution, Keep products and materials in use, Regenerate natural systems

(continued)

Table 5.8 (continued)

Circulytics—indicator category 2. Outcomes (*Source* EMF 2022)		
Theme	General	Indicator
8. Plant, property, and equipment assets	Are you procuring and decommissioning your plant, property, and equipment assets in ways that support a circular economy?	8a. What % (by mass) of your plant, property, and equipment assets procured in the financial year were procured with the following circular procurement approaches?
		8b. Does your company have policies or agreements in place for the end-of-use of existing plant, property and equipment assets (all assets) that enable recirculation in practice?
9. Water	If you operate in a water-intensive industry, are you using water in a circular way?	9a. What % (by volume) of your annual water demand is from each of the following sources: Precipitation harvesting; Cascading use of water (direct use of untreated wastewater, in a manner that is safe for the environment and human health); Internally recirculated water; Seawater; Non-potable water from freshwater areas that are not classified as water-stressed?
		9b. Which % (by volume) of your water withdrawal have you reviewed for SMART reduction targets?
		9c. To what extent do you have plans in place to extract surplus nutrients, metals, chemicals, heat and similar valuable resources before discharging the water used in your processes and operations? 9c (continued). With processes in place to extract surplus nutrients, metals, chemicals, heat and similar valuable resources from water used in operations are the majority of the extracted resources subsequently recirculated (e.g. through heat exchange, as nutrient recirculation that meets the qualifying conditions, etc.)?
		9d. What % (by volume) of water annually used in your operations leaves your infrastructure for reuse elsewhere (as part of symbiosis, cascading), etc.?
10. Energy	Are you procuring renewable energy and (if you are an energy provider) producing renewable energy to support a circular economy?	10a. What % of energy (electricity, heat, and fuel) for your operations is renewable energy?
		10b. What % of the energy you produce is renewable energy?

(continued)

Table 5.8 (continued)

Circulytics—indicator category 2. Outcomes (*Source* EMF 2022)		
Theme	General	Indicator
11. Finance	If you are a financial institution, are you intentionally financing companies and projects that support a circular economy?	11a. What % of each of the following categories' total in USD do you screen positively for circular economy alignment? Lending; Fixed Income; Private Equity; Listed Equity; Other (specify)
		11b. What % of each of the following categories' total in USD goes toward financing the circular economy? Lending; Fixed Income; Private Equity; Listed Equity; Other (specify)

required to perform a self-assessment of their circular performance. In particular, the CTI "*focuses primarily on the circular and linear mass that flows through the company, in which design, procurement and recovery models are crucial levers to determine how well a company performs. In addition to the ability to close the loop, CTI provides insights into overall resource use optimization and the link between the company's circular material flows and its business performance. The framework does not evaluate absolute environmental and social impacts. However, it gives insights into how circularity helps to achieve sustainability objectives related to climate and nature. This shows the circular economy as a key enabler in reaching sustainability objectives*" (WBCSD 2023). The latest version of the CTI is the V4.0 (May 2023). In particular, the tool is based on the assessment of the material flows of the company (Fig. 5.5) and it is organized according to four main topics, namely: Close the Loop, Optimize the Loop, Value the Loop, Impact of the Loop. For each topic a set of quantitative indicators is provided, as shown in Table 5.9.[10]

The CTI can be applied by manufacturers at the product level, by facility managers at the building level and by maintenance and remanufacturing operators at the process level of re-strategies (e.g. repair, reuse, remanufacture, repurpose, etc.) to assess their material flows and improve circularity performance. However, although clear in its intent, the application of the CTI could be challenging especially for small and medium building companies due to its quantitative nature. Quantitative CT Indicators require a considerable "information need" and a consequent significant availability of punctual data. This aspect opens to the issue—discussed in the next paragraph—of "support information tools" (including ICT solutions) that are becoming more and more necessary for the collection of data required for the calculation of the quantitative indicators, thus for performing the circularity assessments. Data sources are multiple in the context of building management therefore it is necessary to rationalize and understand which data are necessary to acquire, which are the most accurate sources and the most appropriate data collection technologies to implement.

---

[10] See "Circular Transition Indicators V4.0. Metrics for business, by business" at https://www.wbcsd.org/contentwbc/download/16345/233646/1 for the complete tool and the quantitative formulas of the Circular Transition Indicators.

Table 5.9 Circular Transition Indicators (CTI) by WBCSD. *Source* WBCSD (2023)

Circular Transition Indicators (CTI)		
Topic	Indicator	Description (*Source* WBCSD 2023)
Close the loop	% material circularity	% material circularity is the weighted average between % circular inflow and % circular outflow. The % circular inflow is determined by the % non-virgin content and % renewable content (sustainably grown bio-based sources). The % circular outflow is determined by the % recovery potential (which is focused on design) and the actual recovery. These three pillars address different aspects of the business: procurement for inflow, design for potential recovery and business model innovation (closed) and legal and partnerships (open) for the actual recovery
	% water circularity	Circularity of water is determined through the % circular water inflow and % circular water outflow, which in turn depend on local water conditions
	% renewable energy	Ratio between renewable energy (annual consumption) and total energy (annual consumption)
Optimize the loop	% critical material	The % critical inflow highlights the share of the inflow considered critical calculated as the ratio between the mass of inflow defined as critical and the total mass of linear inflow. Companies can refer to internal critical materials lists or the European Commission list of Critical Raw Materials (CRMs) available at: https://single-market-economy.ec.europa.eu/sectors/raw-materials/areas-specific-interest/critical-raw-materials_en
	% recovery type	% recovery type focuses on how the company recovers outflow and recirculates it into the value chain. Recovery type is applied to % actual recovery. The results provide a breakdown of the recovered outflow in shares reused/repaired, refurbished, remanufactured recycled or biodegraded
	Actual lifetime	Ratio between product actual lifetime and average product actual lifetime. This means the duration of life that the product actually experiences, on average, rather than design life or warranty period. The actual lifetime indicator provides a higher score for products that stay in use for longer than the industry average
Value the loop	Circular material productivity	Companies can calculate circular material productivity by dividing revenues generated by the mass of linear inflow as considered in the Close the Loop module. Ratio between revenue and total mass of linear inflow. The greater the circular material productivity, the better a company is at decoupling financial performance from linear resource consumption

(continued)

Table 5.9 (continued)

Circular Transition Indicators (CTI)		
Topic	Indicator	Description (*Source* WBCSD 2023)
	CTI revenue	A company's CTI revenue is its revenue adjusted for the % material circularity of its product portfolio. Using the Close the Loop results, a company measures its circular CTI revenue by multiplying the sum of a product (group) or business unit's weighted average of the % circular inflow and % circular outflow and multiplying that by the revenue generated by that product (group) or business unit. As outlined under Close the Loop, calculate both % circular inflow and % circular outflow based on weight of the material flows
Impact of the loop	GHG impact	Greenhouse gas (GHG) impact aims to provide companies with a high-level indication of the GHG emissions savings they may obtain by applying circular strategies. For the material inflow impact on GHG emissions measure the savings of sourcing a higher percentage of recycled content for technical materials; as well as the impact of renewable (sustainably grown) biobased materials versus non-renewable (conventionally grown) biobased materials on GHG emissions. For the GHG impact of circular outflow, CTI v4.0 focuses on the difference in the impact on the material carbon footprint of enabling higher value retention recovery (reuse, refurbish, remanufacture) and recycling, versus linear disposal methods (landfill, incineration)
	Nature impact	This indicator provides an initial screening of land-use impacts from the material extraction and cultivation related to a company's material inflow. It helps companies understand how their circular performance impacts nature by measuring the land-use impacts of their current inflow and potential improvement by shifting to circular sourcing. The indicator is especially relevant for companies that are highly dependent on materials. There are three dimensions to consider when estimating the impact of land-use change on nature: the extent of land use (estimated physical extent of land transformed or occupied, based on volume of raw commodity and yield data), the condition of the land used (estimated change in condition of land relative to intact state, based on practices) and the significance of the land (estimated global significance of land for global biodiversity, based on its importance for threatened species). Formula: extent x condition x significance

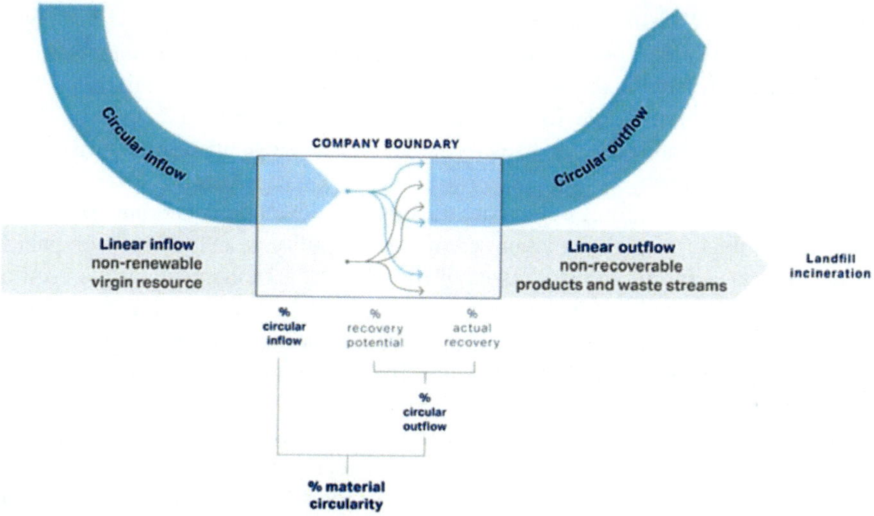

Fig. 5.5 Material flows assessed by the CTI by WBCSD. *Source* WBCSD 2023

5.3 Reporting Circular Practices Within Building Management and Related Performance: Emerging Needs and Game Changers

The topic of sustainability assessment is currently at the centre of the debates by European and international regulatory bodies. What is certain is that the sustainability assessment, as also stated by the recent Corporate Sustainability Reporting Directive (CSRD) and the EU Taxonomy Regulation, can no longer neglect the inclusion of issues relating to the circular economy and circularity which are increasingly becoming central and priority aspects towards the achievement of the key objectives of the European Green Deal and the UN Agenda 2030 for Sustainable Development.

In the building and Real Estate sector, the new circularity requirements under the EU Taxonomy and the CSRD probably will initially affect investor entities and mortgage providers, such as banks, insurers, institutional investors, pension funds, etc. who first fall within the scope of the CSRD. However, these subjects will make investment choices oriented towards sustainability and circularity (e.g. green investments) that concern the different sector operators involved in performing the economic activities listed by the EU Taxonomy, namely "new constructions", "renovations" and "buying and selling of buildings". Therefore, all the stakeholders operating in the sector will be consequently rapidly involved in the sustainability reporting activities. Indeed, investors could enquire project promoters whether the building meets the technical criteria of the EU Taxonomy before and after the building completion (new building or renovation) and the project developers in turn will have to provide evidence to the bank when applying for green funding. Sustainability assessments, in this sense, will be aimed at assessing the project according to two levels: (i) characteristics of the asset (both for new projects and for buildings to be renovated); (ii) characteristics related to the operation, management and use of the building. Therefore, also the operators directly responsible for the design and management of the buildings will be indirectly involved in the preliminary sustainability assessment of the investment project.

In particular, for building companies (e.g. project development companies, design firms, manufacturers, construction companies, facility management companies, maintenance operators, etc.) addressed by the CSRD it is fundamental to start setting the ground to (i) perform a double materiality (inward and outward perspective) assessment of ESG issues and (ii) set up structured monitoring and reporting procedures in order to keep track of ESG goals over time and accordingly adjust and improve strategies.

The transition process will not be free of efforts, especially for small operators—not yet addressed by the CSRD but in a few years—that need to improve their preparedness for meeting the new sustainability reporting regulations. In this regard, voluntary support tools—such as the guidelines developed by national and European standardization bodies, including the fundamental standards of the ISO 59000 series, together with the initiatives of associations such as the Ellen MacArthur Foundation (EMF) and others, are providing effective support to companies.

Based on the review introduced in the previous paragraphs, beside metrics (clearly defined), it is possible to identify some key areas of building practices that require improvements towards circularity, articulated on two levels: strategic and operational. For each of the identified area, the current emerging needs as well as the related potential "game changers" for companies are outlined.

In particular, at the strategic level:

- *Governance*

A crucial factor for the success of the transition process towards green and circular practices is the commitment of the company's board, which must first adopt an approach aimed at creating value in the medium-long term, combining the results of the financial statements with the caused environmental and social impacts, according to a logic aimed at integrating the sustainability objectives into the business plan and the quality and control systems. However, if on the one hand the board—by defining orientation lines and priorities—represents the essential starting point and the engine of this process, on the other hand the achievement of results depends on the engagement of all the levels of responsibility within the company. Therefore, it will be a primary concern of the board to set information and training actions in order to stimulate the engagement at different company levels, by sharing visions and objectives with all its staff towards the achievement of common objectives. In this whole process, a long-term vision by the company board on circularity benefits is fundamental. Indeed, the effort for complying with the new CSRD requirements may lead in the short term to an increase of costs and a decrease of profit margins. However, the effort will be beneficial also to increase the ability of the company itself to attract investment capitals. Indeed, since the market of the Real Estates is mostly stimulated by investments and funding from investors and financial supporters that in turn will be subjected to the CSRD, the integration of sustainability strategy and the identification of sustainability risks and opportunities will become almost a prerequisite for attracting capital and gaining a competitive advantage.

- *Integration*

This issue mainly refers to the need for: (i) integration of the circular economy principles within the company operations; (ii) integration of sustainability within performance management systems (iii) integration of the environmental dimensions in the practices of monitoring and controlling of the performance and impacts of the company; (iv) integration of the reporting of outputs, i.e., the results of company core activities, with the reporting of outcomes, i.e., the effects and impacts generated on the environment and on all the stakeholders and the community in the short-, medium- and long-term; (v) integration of the measurement of the environmental impacts of the company with those generated by the other actors of the company's value chain/supply chain. The integration of the three dimensions of sustainability in the reporting will lead to the drafting of a complete report which, in addition to ensure compliance of contents with regulatory requirements, makes it possibile to: (i) identify possible weakness areas and promptly adjust strategies to manage and mitigate risk, thus to compensate and reduce the negative impacts of the company on the social and environmental dimensions; (ii) capitalize the ESG performance of the company in competitive and reputational terms and at the same time; (iii) use the sustainability report itself as a communication vehicle for informing and engage stakeholders and, ultimately, gaining insights on marketing accountability.

While, at the operational level:

- *Data Management*

Often data concerning company operations are unstructured and not collected according to standardized procedures and shared information management tools. This could hinder the ability of companies of organizing and storing data overtime, thus reducing the availability of ESG data needed to draft the sustainability reports. Hence, it is important to set a structured and systematic approach to data management, shared with all the levels and roles of the company. In particular, it is advisable to:

(i) understand which information is useful (therefore a priority) to calculate the ESRS (Table 5.4) (or the indicators of other voluntary tools for those companies not yet under the obligation of the CSRD, Tables 5.6, 5.7, 5.8 and 5.9) and understand what are the related possible data sources (Fig. 5.6);
(ii) identify the most suitable data sources by balancing data accuracy and ease of data detection/collection;
(iii) define standardized procedures for the collection and management of data, information and documents concerning the various aspects (administrative, technical, economic, etc.) of the building and its parts. This step involves the

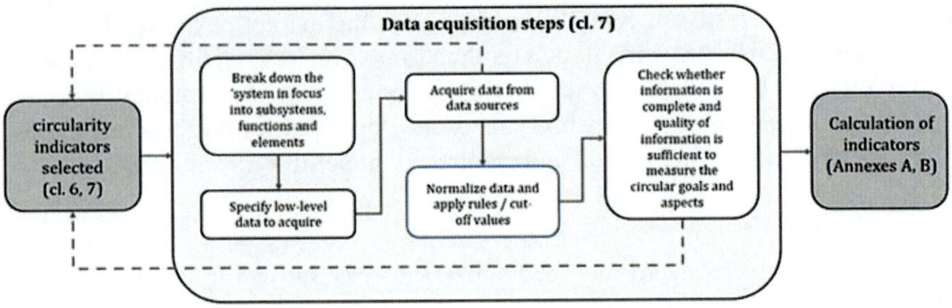

Fig. 5.6 Data acquisition process for the calculation of circularity indicators. *Source* ISO/DIS 59020:2023

adoption of a "registry system"[11] based on a system for the classification and coding of the building elements, e,g. Onmiclass by CSI (see Chap. 3) and definition of common templates, procedures and measurement tools for performing an "inventory process"[12];

(iv) adopt an information system capable of acquiring, store (database) and process the collected data, setting the stage for the drafting of the sustainability reports.

– *Digital Information and Communication Technology (ICT) Solutions*

It is advisable to implement a digital information system (or information platform) with functions dedicated to "digital reporting". The latter will allow the company to communicate, internally and externally, its social, economic and environmental performance in a smart, dynamic and interactive way. The digital information system can support the company in meeting the CSRD request to track risks and impacts along the entire value-chain. In fact, the creation of a structured and accessible information base can support: (i) mutual learning between the company and its stakeholders in a logic of continuous improvement and (ii) sharing of ESG data and information with internal and external stakeholders, establishing a transparent system for reporting, based on co-responsibility among the value-chain stakeholders for the achievement of sustainable development goals.

[11] A "registry system" consists of a framework of criteria useful for the classification and coding of spatial and technical elements of a building. It also includes an apparatus of standardized templates useful for collecting information, over time according to the same formalized scheme.

[12] An "inventory process" can be defined as a gradual process of acquisition of data and information on a Real Estate asset concerning different aspects (e.g. dimensional, physical, technical, administrative, performance, financial, legal, etc.). The inventory process can involve a plurality of integrated activities such as analysis, audits, surveys, collection of technical, administrative and legal data and documents, etc.

References

EFRAG (2022a) [Draft] ESRS E5 resource use and circular economy. https://www.efrag.org/Assets/Download?assetUrl=%2Fsites%2Fwebpublishing%2FSiteAssets%2F12%2520Draft%2520ESRS%2520E5%2520Resource%2520use%2520and%2520circular%2520economy.pdf. Accessed July 2023

EFRAG (2022b) [Draft] European sustainability reporting guidelines 1 double materiality conceptual guidelines for standard-setting. https://www.efrag.org/Assets/Download?assetUrl=/sites/webpublishing/SiteAssets/Appendix%202.6%20-%20WP%20on%20draft%20ESRG%201.pdf. Accessed July 2023

EMF (2022) Circulytics. Indicators. https://emf.thirdlight.com/link/1pzbxosbi6hl-ei3tq6/@/#id=2. Accessed July 2023

EMF (2023) Measure business circularity: circulytics. https://ellenmacarthurfoundation.org/resources/circulytics/overview. Accessed July 2023

Ernst & Young - EY (2022) Corporate sustainability reporting directive. https://assets.ey.com/content/dam/ey-sites/ey-com/en_gl/topics/assurance/assurance-pdfs/ey-corporate-sustainability-reporting-directive-brochure-june-2022.pdf?download. Accessed March 2023

Ernst & Young - EY (2023) Sustainable finance disclosure regulation: getting ready for "level II" application. https://www.ey.com/en_lu/sustainability-financial-services/sustainable-finance-disclosure-regulation--getting-ready-for--le#:~:text=Its%20purpose%20is%20to%20enable,widely%20accepted%20by%20the%20market). Accessed July 2023

EU (2023a) EU taxonomy navigator. Construction and real estate. https://ec.europa.eu/sustainable-finance-taxonomy/sectors/sector/7/view. Accessed July 2023

EU (2023b) EU taxonomy navigator. 2. EU taxonomy calculator—a set-by-step guide on reporting obligations. https://ec.europa.eu/sustainable-finance-taxonomy/home. Accessed July 2023

Eurostat (2016) Key figures on Europe (pp. 161–164). https://ec.europa.eu/eurostat/documents/3217494/7827738/KS-EI-16-001-EN-N.pdf/bbb5af7e-2b21-45d6-8358-9e130c8668ab. Accessed July 2023

WBCSD (2023) Circular transition indicators V4.0. Metrics for business, by business. https://www.wbcsd.org/contentwbc/download/16345/233646/1. Accessed July 2023

Standards and Laws

EU (2023c) Annex II of "Commission Delegated Regulation (EU) [...] of 27.6.2023 supplementing Regulation (EU) 2020/852 of the European Parliament and of the Council [...]". https://finance.ec.europa.eu/system/files/2023-06/taxonomy-regulation-delegated-act-2022-environmental-annex-2_en_0.pdf. Accessed July 2023

ISO/DIS 59020:2023 Circular economy. Measuring and assessing circularity

Chapter 6
Conclusions

The recent updates of the legislation on circular economy, circularity and, more generally, environmental sustainability at the European scale demonstrate the strong interest by EU to promote the application within the building sector of new circular models based on product life-extension strategies (e.g. reuse, repair, remanufacturing, etc.) and service-based supply formulas (e.g. product-as-a-service, sharing of product use, renting for multiple product uses, etc.) according to a "resource-resource" approach aimed at maintaining the value embedded in building elements as long as possible, thus going beyond the linear "take-make-dispose" models.

In particular, the new obligations and policies introduced by EU in the context of the Circular Economy Action Plan (CEAP)—including Sustainable Finance Disclosure Regulation (SFDR), EU Taxonomy Regulation, Corporate Sustainability Reporting Directive (CSRD), Level(s), etc.—are leading sector operators to face a sudden change of practices. The latter requires a holistic approach since it encompasses all the application scales (organization; building; process; product), all the decision-making levels (strategic; tactical; operational), all the interrelated building phases (design; construction; use and management; end-of-life) and all the building parties (project developers, promoters and investors; design firms; product manufacturers; construction companies; property and building managers; providers of facility management services; demolition operators; waste managers; sustainability consultants, etc.). Consequently, building stakeholders need to rapidly develop new knowledge bases and to acquire new skills useful for understanding: (i) how to best integrate circularity into their business models, (ii) how to evaluate the viability of circular strategies, (iii) how to update their operational procedures and tools and (iv) how to acquire the physical (e.g. machinery, warehouses, etc.), technological (e.g. digital information system or platform for data management, ICT solutions for remote monitoring, etc.) and relational (e.g. new supply segments to integrate into the value chain, creation of new collaborative networks, etc.) infrastructures necessary for the implementation of circular practices. However, the novelty and complexity of the issue implies upstream harmonization efforts, which requires time for a careful

N. Atta, *Green Approaches in Building Design and Management Practices*,
Digital Innovations in Architecture, Engineering and Construction,
https://doi.org/10.1007/978-3-031-46760-8_6

collective review of basic references and traditional practices (methods and tools) of building design, management and performance assessment in light of circularity approaches.

In this regard, the book contributes to support building stakeholders in the integration of circularity within their building practices by proposing:

- new circularity-oriented approaches to the development of Briefing Documents (BDs) for the design of circular buildings, including new circularity-related BDs contents (Chap. 3);
- new contents for expressing circularity-related requests within Invitations to Tender (ITTs) for building products and Facility Management (FM) services provision (Chap. 4);
- new approaches and tools for reporting circularity performance of building practices (Chap. 5).

Based on the analysis conducted on the complex apparatus of mandatory and voluntary support tools on a European and international scale, the book identifies some cornerstones of circular transition for a widespread adoption of circular practices in the building and Real Estate sector, which represent at the same time key issues to further explore in the next researches towards a Circular Building Process. In particular:

- *Design and Planning of material flows.* Definition of: (i) type of input and output materials/products, (ii) balance of volumes of the input and output materials/products, (iii) demand and destination of output materials/products. These aspects must be clarified upstream in order to allow the effective and efficient implementation of reuse practices of materials/products.
- *Setting the stage for "servitization" models.* Service-oriented models are based on product access (use) instead of product ownership. This shift in supply strategy involves significant infrastructural changes, both in terms of technology (e.g. real-time monitoring of use, smart billing solutions, etc.), but also and above all in terms of partnerships between stakeholders. The interactions and contractual relationships need to be redefined in a win–win perspective by reallocating roles and functions (even new ones) to the involved operators.
- *Definition of logistic aspects.* Logistics assumes a significant relevance in circular models of building elements reuse. If not well planned, logistic aspects could lead to process inefficiencies. Hence, to design and plan logistics, it is necessary for instance to preliminarily: (i) assess and estimate the need for new warehouses for product storage and their geographical location, as well as the need for operators capable of collecting and distributing the products; (ii) implement assessments of the impacts of logistics and the related economic sustainability.
- *Strategic design of the stakeholder network.* Based on the specificity of the circular practices to implement, companies must identify which operators (e.g. production, remanufacturing, logistics and distribution companies, etc.) they need to include into their value-chain. These operators may not already be part of the current company supply chain or part of the company core business, however they need to be included in the value network to allow the implementation of circular

practices. In this regard, the interests of the various parties must be understood and carefully taken into consideration to establish win–win relationships. Such a network shares complementary resources and competences, reaching the critical mass and robustness necessary to initiate circular practices. Morevoer, it also serves as an innovation hub for circular experimentations at product and process levels.

- *Implementation of ICT infrastructures and digital technologies.* An information system (or information platform) for the advanced collection and management of data, information and documents is necessary starting from the building design phase up to the management phase. Among its main support functions for circularity, it is possible to mention: (a) unique classification and coding of building elements. The building is broken down into systems and elements (also represented graphically in 2D or 3D), which become "information collectors", overtime creating the central database; (b) collection, archiving and processing of static and dynamic data concerning the various aspects of the building (e.g. technical, contractual, financial, environmental, performance, administrative, etc.); (c) connection in a single communication environment of different stakeholders (data holders) of the value network to facilitate information collection and exchange, communication and collaboration at both strategic and operational level; (d) implementation of sensor networks and blockchain technologies to ensure the traceability of the building elements and to update their chains of custody over time; (e) document management, i.e. development and archiving of documents such as product passports, disassembly plans, maintenance plans, waste management plans, performance assessment reports, etc.; (f) creation of digital twins for the analysis of the propensity-to-circularity of the elements, also carrying out simulations of re-strategies (including disassembly, remanufacturing, repurposing, etc.) useful for the related feasibility and sustainability assessments; (g) maintenance management of building elements for the extension of their useful life. Through the implementation of sensing technologies and Internet of Things (IoT) solutions, maintenance activities are planned and scheduled according to the profiles of use and actual conditions of the building elements; (h) management of service-based models (e.g. product-as-a-service, renting, etc.) and related payment methods (e.g. pay-per-use, pay-per-period, etc.) which require the implementation of a real-time usage and consumption monitoring system as well as a smart billing system, based on sensing technologies and IoT solutions; (i) management of SLA&KPI system (based on updated regulatory requirements) to evaluate the circularity performance of building practices and develop sustainability reports.

In light of the complexity inherent in the transition process towards circular building practices, to achieve Circular Economy objectives it is necessary to adopt a gradual approach to the integration of circularity. It is advisable for sector operators to gradually set goals that are achievable but increasingly ambitious, to be pursued through the solid collaboration with the different stakeholders of the value network, exploiting the complementarity of skills and resources, and with the support of a robust (upcoming) sector regulatory framework on circularity.

Glossary

Accessibility Ability for ease of access to components for disassembly, refurbishment, replacement, or upgrade (ISO 20887:2020)

Adaptability Ability to be changed or modified to make suitable for a particular purpose (ISO 6707-1:2017)

Add value Process of increasing the value of the object of consideration (i.e., a resource) (ISO/DIS 59004:2023)

Agreement Statement agreed between the demand organization and the provider of services or products (ISO 41011:2017)

Assembly Set of related components attached to each other (ISO 6707-1:2017)

Asset register Collection of records holding information about facility assets in terms of their manufacturer, vendor, make, model, specifications, date of acquisition, initial cost, maintenance cost and requirements, accumulated depreciation and written-down value (BS 8587:2012)

Asset Item, thing or entity that has potential or actual value to an organization (ISO 55000:2014). Anything considered by an organisation as having a positive value, especially financial (BS EN 15221-2:2006)

Availability Property of being accessible and usable upon demand by an authorized entity (ISO/IEC 27000:2018)

Benchmark Reference point or metric against which a strategy, process, performance and/or other entity can be measured (BS EN 15221-7:2012)

Benchmarking Process of comparing strategies, processes, performances and/or other entities against practices of the same nature, under the same circumstances and with similar measures (BS EN 15221-7:2012)

Big Data Data set(s) with characteristics (e.g. volume, velocity, variety, variability, veracity, etc.) that for a particular problem domain at a given point in time cannot be efficiently processed using current/existing/established/traditional technologies and techniques in order to extract value (ISO/IEC JTC 1 - Information Technology. Big Data. Preliminary Report 2014)

N. Atta, *Green Approaches in Building Design and Management Practices*, Digital Innovations in Architecture, Engineering and Construction,
https://doi.org/10.1007/978-3-031-46760-8

Building Information Modeling (BIM) Process of designing, constructing or operating a building or infrastructure asset using electronic object-oriented information (PAS 1192-2:2013)

Built environment Physical construction result intended to serve a function or user activity. Note: The built environment may be viewed as a system of either built space or built structure (BS EN ISO 12006-2:2020)

Business model Organization's chosen system of interconnected and interdependent decisions and activities that determines how it creates, delivers and captures value over the short, medium and long term Note 1: a value creation model involves external processes (e.g. transportation, take-back, etc.) beyond those of the organization's processes (e.g. education, financing, etc.) and the solutions it provides. Note 2: a business model is a subset of value creation models wherein the chosen system determines how the organization creates, delivers, and captures economic value (ISO/DIS 59010:2023)

Certification Third-party attestation related to an object of conformity assessment, with the exception of accreditation (ISO/IEC 17000:2020, ISO 22095:2020)

Chain of custody Process by which inputs and outputs and associated information are transferred, monitored and controlled as they move through each step in the relevant supply chain (ISO 22095:2020)

Chain of custody system Set of measures designed to implement a chain of custody, including documentation of these measures. Note 1: the purpose of a chain of custody system is to provide credibility that the given material or product has a set of specified characteristics. Note 2: the information linked to materials or products is transferred, monitored and controlled throughout the entire supply chain or parts of it (ISO 22095:2020)

Circular economy Economy that is restorative and regenerative by design, and which aims to keep products, components and materials at their highest utility and value at all times, distinguishing between technical and biological cycles (ISO 20400:2017). Economic system that uses a systemic approach to maintain a circular flow of resources, by recovering, retaining or adding to their value, while contributing to sustainable development. Note 1: Resources can be considered concerning both stocks and flows. Note 2: From a sustainable development perspective, the inflow of virgin resources is kept as low as possible, and the circular flow of resources is kept as closed as possible to minimize emissions and losses (waste) (of resources) from the economic system (ISO/DIS 59004:2023)

Circular flow of resources Systematic cycling of the provision and use of resources within technical or biological cycles. Note 1: resources can be considered concerning both stocks and flows. Note 2: the biological and technical cycles represent loops into the complex system of resource flows in the economy (ISO/DIS 59004:2023)

Circularity assessment Evaluation and interpretation of results and impacts from a circularity measurement. Note 1: assessment considers, for example, data from resource flow measurements, environmental, economic and social impacts as well as suitable circularity indicators used to monitor progress of circular goals and actions. Note 2: assessment includes consideration of the circularity performance and other complementary methods such as life cycle assessment (ISO/DIS 59020:2023)

Circularity indicator Metric used to measure one or more circularity aspects. Note: a circularity indicator can represent a measurable aspect or combination of aspects of a resource, a process or action (ISO/DIS 59020:2023)

Circularity measurement Process to help determine the circularity performance through collection, calculation or compilation of data or information (ISO/DIS 59020:2023)

Circularity performance Degree to which a set of circularity aspects align with the principles for a circular economy (ISO/DIS 59020:2023)

Classification Systematic identification and arrangement of business activities and/or records into categories according to logically structured conventions, methods, and procedural rules represented in a classification system (ISO 15489-1:2004). System for grouping and categorising items with similar characteristics (attributes) (BS EN 15221-4:2011)

Client Organisation that procures facility services by means of a Facility Management agreement. Note: The client acts on a strategic level and has a general and/or key function in all stages of the relationship with the service provider. The customer specifies the facility services (BS EN 15221-1:2006)

Closed loop system System by which products or resources are used and then recovered and turned into new products or recovered resources, without losing their inherent properties (ISO/DIS 59004:2023)

Cloud service One or more capabilities offered via cloud computing invoked using a defined interface (ISO/IEC 17788:2014)

Competence Ability to apply knowledge and skills to achieve intended results (ISO/IEC 27000:2018)

Compliance obligations Legal requirements that an organization has to comply with and other requirements that an organization has to or chooses to comply with. Note 1: Compliance obligations are related to the environmental management system. Note 2: Compliance obligations can arise from mandatory requirements, such as applicable laws and regulations, or voluntary commitments, such as organizational and industry standards, contractual relationships, codes of practice and agreements with community groups or non-governmental organizations (ISO 14001:2015)

Component Product manufactured as a distinct unit to serve a specific function or functions. Note 1: Components can be manufactured, prefabricated, or built or formed on site, and can be basic or complex units. Note 2: a complex unit can also be considered an assembly (ISO 20887:2020)

Conditions of contract Terms that collectively describe the rights and obligations of contracting parties and the agreed procedures for the administration of their contract (ISO 10845-1:2020)

Conformity Fulfillment of a requirement ((ISO/IEC 27000:2018)

Conformity assessment Demonstration that specified requirements are fulfilled. Note 1: conformity assessment can be performed as first-party activity, second-party activity or third-party activity (ISO 22095:2020)

Constructed asset Anything of value that is constructed or results from construction operations (ISO 15686-1:2011)

Construction product Product intended to be used as a construction resource. Note: Construction products have different complexity and can, by themselves or together with others, make up the parts in any level of assembly of construction entities (BS EN ISO 12006-2:2020)

Continual improvement Recurring activity to enhance performance. Note 1: Enhancing performance relates to the use of the environmental management system to enhance environmental performance consistent with the organization's environmental policy. Note 2: The activity need not take place in all areas simultaneously, or without interruption (ISO 14001:2015). The process of establishing objectives and finding opportunities for improvement is a continual process through the use of audit findings and audit conclusions, analysis of data, management reviews or other means and generally leads to corrective action or preventive action (ISO 9000:2015)

Contract data Document that identifies the applicable conditions of a contract and states the associated contract-specific data (ISO 10845-1:2020)

Contract Agreement under which two parties undertake to exchange a product or a service for a payment (ISO 41011:2017). Binding agreement (ISO 9000:2015)

Control Comparison of actual performance with planned performance, analysing variances and taking appropriate corrective and preventive action as needed (BS ISO 21500:2012)

Customer Organisational unit that specifies and orders the delivery of facility services within the terms and conditions of a Facility Management agreement. Note The customer acts on a tactical level (BS EN 15221-1:2006)

Deliverable Measurable and verifiable outcome, result or item to be produced within a specific timeframe to complete a project or part of a project (ISO 22128:2008)

Demand Stated requirement for services or products to be delivered (ISO 41011:2017)

Design for disassembly Approach to the design of a product or constructed asset that facilitates disassembly at the end of its useful life, in such a way that enables components and parts to be reused, recycled, recovered for energy or, in some other way, diverted from the waste stream (ISO 14021:2016)

Device With regard to the Internet of things, this is a piece of equipment with the mandatory capabilities of communication and the optional capabilities of sensing, actuation, data capture, data storage and data processing (ITU-T Y.2060)

Disassembly Process whereby a product is taken apart in such a way that it could subsequently be reassembled and made operational (EN 45553:2020). Non-destructive taking-apart of a construction works or constructed asset into constituent materials or components. Note: this process can be applied to a product, module, system, component, or assembly (ISO 15392: 2019)

Disassembly step Operation that ends with the removal of a part (EN17902:2022)

Document Physical support of the information in a specific form. This may take the form of a paper sheet, the screen of a video monitor of a computer system, an electronic board, a blackboard, etc. and the figures, type, size and distribution on the available surface may vary without affecting the main purpose of the information system. Note: A document is permanent. Program results displayed on a screen do not make any document unless it is stored. Document can be information stored in a database which can be shown on a screen or printed out (BS EN 13460:2009). Fixed and structured amount of information that can be managed and interchanged as a unit between users and systems (ISO 29845:2011)

Documents review Activity aimed at searching, selecting, analysing, gathering and organizing the various and heterogeneous documents coming from design, construction, operations and maintenance phases

Due diligence Process through which organizations proactively identify, assess, prevent, mitigate and account for how they address their actual and potential adverse impacts as an integral part of decision-making and risk management (ISO 20400:2017). Process of conducting a walkthrough survey and appropriate inquiries into the physical condition of a commercial real estate's improvements, usually in connection with a commercial real estate transaction. The degree and type of such survey and inquiry may vary for different properties, different user purposes, and time allotted (ASTM E 2018-08)

Durability Ability of a constructed asset or any of its components to perform its required functions in its service environment over a specified period of time without unforeseen maintenance or repair. Note: preventive or routine maintenance are foreseen measures intended to increase functional service life (ISO 20887:2020)

Ecodesign Design and development based on life cycle thinking aimed at supporting sustainable development. Note: other terminology used includes "Environmentally Conscious Design (ECD)", "Design for Environment (DfE)", "green design" and "environmentally sustainable design" (ISO 14006:2020; ISO/DIS 59004:2023)

Element Construction entity part which, in itself or in combination with other such parts, fulfils a predominating function of the construction entity (i.e. elements are: external wall, floor, roof, etc.) (e.g. predominating functions are: space enclosing, supporting, servicing, furnishing) (ISO 12006-2:2015)

Employer Person or organization intending to or entering into a contract with the contractor to supply goods, carry out construction works and/or provide services. Note: The definitions for "employer" and "contractor" are such that relevant subclauses of this document can be applied at any point in the supply chain. For example, in a contract, a "contractor" can be an "employer" and a "contractor" can be a "subcontractor" (ISO 10845-1:2020)

End user Person receiving facility services. Note: A visitor could also be an end user (BS EN 15221-1:2006)

Energy Management System (EMS) Equipment/infrastructure used to monitor, measure and control the energy consumption in private households, residential buildings or industrial customer installations. Note 1: The term EMS is also commonly used to refer to a set of applications used by operators of transmission power grid to monitor, control, and optimize the performance of the generation and/or transmission system (ISO/IEC 27019:2017)

Engagement Involvement in, and contribution to, activities to achieve shared objectives (ISO 9000:2015)

Environment Surroundings in which an organization operates, including air, water, land, natural resources, flora, fauna, humans, and their interrelationships. Note 1: surroundings can be described in terms of biodiversity, ecosystems, climate or other characteristics (ISO 14001:2015; ISO/DIS 59004:2023)

Environmental impact Any change to the environment, whether adverse or beneficial, wholly or partially resulting from any element of an organization's activities, products or services that can interact with the environment (ISO 14001:2015; ISO/DIS 59004:2023)

Environmental Management System (EMS) Part of the management system used to manage environmental aspects, fulfil compliance obligations, and address risks and opportunities (ISO 14001:2015)

Environmental policy Intentions and direction of an organization related to environmental performance, as formally expressed by its top management (ISO 14001:2015)

Ethical Behaviour Behaviour that is in accordance with accepted principles of right or good conduct in the context of a particular situation and is consistent with international norms of behaviour (ISO 26000:2010)

Event Occurrence or change of a particular set of circumstances. Note 1: An event can have one or more occurrences, and can have several causes and several consequences. Note 2: An event can also be something that is expected which does not happen, or something that is not expected which does happen. Note 3: An event can be a risk source (ISO 31000:2018)

Extended Producer Responsibility (EPR) Producer's liability for a product extended to the treatment or disposal of post-consumer products (ISO/DIS 59010:2023)

Facility Management (FM) Organizational function which integrates people, place and process within the built environment with the purpose of improving the quality of life of people and the productivity of the core business (ISO 41011:2017). Integration of processes within an organisation to maintain and develop the agreed services which support and improve the effectiveness of its primary activities (BS EN 15221-1:2006)

Facility management agreement Written or oral agreement stating the terms and conditions for provision of facility services between a client and an internal or external service provider (BS EN 15221-1:2006)

Facility management service provider Organisation that provides the client with a cohesive range of facility services within the terms and conditions of a Facility Management agreement. Note: A Facility Management service provider can be internal or external to the client (BS EN 15221-1:2006)

Facility manager Person responsible for the facility management organisation who is the single point of contact for the client on strategic level; leads the FM organisation, ensures quality and continuous improvement and conducts strategic projects and tasks (BS EN 15221-4:2011)

Facility service Support provision to the primary activities of an organisation, delivered by an internal or external provider. Note: Facility services are services related to Space and Infrastructure and to People and Organisation (BS EN 15221-1:2006)

Facility Collection of assets which is built, installed or established to serve an entity's needs (ISO 41011:2017)

Framework Documented set of guidelines to create a common understanding of the ways of working (ISO 37500:2014)

Hierarchy Structure of levels in which each level includes its lower levels. Note: Taxonomies are frequently arranged in a hierarchical structure. Typically, they are related by supertype-subtype, also called parent–child relationships (BS EN 15221-4:2011)

Indicator Measure that provides an estimate or evaluation of specified attributes derived from an analytical model with respect to defined information needs (ISO/IEC 27000:2016)

Information exchange Structured collection of information at one of a number of pre-defined stages of a project with defined format and fidelity (PAS 1192-2:2013)

Information management Task and procedures applied to inputting, processing and generation activities to ensure accuracy and integrity of information (PAS 1192-2:2013). Processing and storage of information in a controlled manner (BS 10008:2008)

Information system Decisional and operational support tool consisting of databases, procedures and functions to collect, store, process, use and update the information necessary for the setting, the implementation and management of the maintenance service (UNI 10951:2001)

Information Representation of data in a formal manner suitable for communication, interpretation or processing by human beings or computer applications (PAS 1192–2:2013)

Infrastructure System of facilities, equipment and services needed for the operation of an organization (ISO 9000:2015)

Interested party or Stakeholder Person or organization that can affect, be affected by, or perceive itself to be affected by a decision or activity. Note 1: to "perceive itself to be affected" means the perception has been made known to the organization. Note 2: the terms "interested party" and "stakeholder" can be used interchangeably (ISO/DIS 59010:2023)

Internet of Things (IoT) Global infrastructure for the information society, enabling advanced services by interconnecting (physical and virtual) things based on existing and evolving interoperable information and communication technologies. Note 1: Through the exploitation of identification, data capture, processing and communication capabilities, the IoT makes full use of things to offer services to all kinds of applications, whilst ensuring that security and privacy requirements are fulfilled. Note 2: From a broader perspective, the IoT can be perceived as a vision with technological and societal implications (ITU-T Y.2060). Infrastructure of interconnected entities, people, systems and information resources together with services which processes and reacts to information from the physical and virtual world (ISO/IEC 20924:2021)

Interoperability Ability of two or more systems to exchange information and to make mutual use of the information that has been exchanged (ISO 5127:2017)

Inventory process A gradual process of acquisition of data and information on a Real Estate asset concerning different aspects (e.g. dimensional, physical, technical, administrative, performance, financial, legal, etc.). The inventory process can involve a plurality of integrated activities such as analysis, audits, surveys, collection and validation of technical, administrative and legal data and documents

Key Performance Indicator (KPI) Measure that provides essential information about the performance (ISO 41011:2017)

Knowledge The result of application, processing, relating, combining of information in specific contexts

Knowledge acquisition Process of locating, collecting, and refining knowledge and converting it into a form that can be further processed by a knowledge-based system (ISO/DIS 37500:2014)

Knowledge base A collection of facts, assumptions, beliefs, and heuristics that are used in combination with a database to achieve desired results, such as a diagnosis, an interpretation, or a solution to a problem (McGraw-Hill Dictionary of Scientific & Technical Terms, 6E, Copyright © 2003 by The McGraw-Hill Companies, Inc.)

Knowledge transfer Structured process of imparting pre-existing or acquired information to a team or a person, to help them attain a required level of proficiency in skill (ISO/DIS 37500:2014)

Life cycle Consecutive and interlinked stages in the life of a solution. Note 1 to entry: the life cycle stages can, e.g. include acquisition of natural resources, design, production, transportation or delivery, use, end-of-use treatment, and end-of-life treatment and disposal. Note 2: within a circular economy, traditional linear life cycle understanding is upended in support of the concept of multiple use cycles. Note 3: examples of end-of-use treatment include recycling, reuse, and remanufacturing (ISO/DIS 59004:2023)

Life cycle approach Consideration of life cycle in decision-making or development processes (ISO 20400:2017)

Life cycle assessment (LCA) Method of measuring and evaluating the environmental impacts associated with a product, system or activity, by describing and assessing the energy and materials used and released to the environment over the life cycle (ISO 15686-5:2017)

Life Cycle Cost (LCC) Methodology for systematic economic evaluation of life-cycle costs over a period of analysis, as defined in the agreed scope Note 1 to entry: Life cycle costing can address a period of analysis that covers the entire life cycle or (a) selected stage(s) or periods of interest thereof (ISO 15686-5:2017). Cost of an asset or its parts throughout its life cycle, while fulfilling the performance requirements (ISO 15686-5:2017)

Life cycle costing Methodology for systematic economic evaluation of life-cycle costs over a period of analysis, as defined in the agreed scope (ISO 15686-5:2017)

Life cycle perspective Consideration of the circularity aspects relevant to a solution during its life cycle and includes environmental, social and economic impacts. Note 1: the main goals of life cycle perspective are to reduce a product's resource use and emissions to the environment as well as improve its socio-economic performance through its life cycle. This can facilitate links between the economic, social, and environmental dimensions within an organization and through its entire value chain. Note 2: in measuring and assessing the circularity performance of a system, a life cycle perspective should be applied. Note 3: this perspective should include all stages of technical or biological cycles over appropriate timescales that are related to that system. Note 4: life cycle thinking can be used as an alternative to life cycle perspective and can be used interchangeably (ISO/DIS 59004:2023)

Linear economy Economic system where resources typically follow the pattern of extraction, production, use and disposal (ISO/DIS 59004:2023)

Materiality Measure of the significance of an element to organizational results (ISO/DIS 59010:2023)

Materiality assessment (analysis) Method to identify and prioritize the issues that are most important to an organization and the stakeholders relevant to its circular economy strategies (ISO/DIS 59010:2023)

Management system Set of interrelated or interacting elements of an organization to establish policies and objectives and processes to achieve those objectives. Note 1: A management system can address a single discipline or several disciplines (e.g. quality, environment, occupational health and safety, energy, financial management). Note 2: The system elements include the organization's structure, roles and responsibilities, planning and operation, performance evaluation and improvement. Note 3: The scope of a management system can include the whole of the organization, specific and identified functions of the organization, specific and identified sections of the organization, or one or more functions across a group of organizations (ISO 14001:2015)

Measurement Process to determine a value (ISO 41011:2017)

Modular Composed of modules for easy construction or arrangement and adaptation or disassembly (ISO 20887:2020)

Module Set of standardized parts or independent units. Note 1: modularization can be key to disassembly in many types of civil engineering works. Note 2: a module could be a type of complex assembly (ISO 20887:2020)

Monitoring Determining the status of a system, a process, a product, a service, or an activity. Note 1: For the determination of the status there can be a need to check, supervise or critically observe. Note 2: Monitoring is generally a determination of the status of an object, carried out at different stages or at different times (ISO 9000:2015)

Obsolescence Loss of ability of an item to perform satisfactorily due to changes in performance requirements (ISO 15686-1:2011)

Organization Person or group of people that has its own functions with responsibilities, authorities and relationships to achieve its objectives. Note: The concept of organization includes, but is not limited to sole-trader, company, corporation, firm, enterprise, authority, partnership, charity or institution, or part or combination thereof, whether incorporated or not, public or private (ISO 14001:2015)

Outsource Make an arrangement where an external organization performs part of an organization's function or process (ISO/IEC 27000:2018)

Outsourcing Business model for the delivery of a product or services to a client by a provider (ISO 37500: 2014)

Outsourcing model Formalized concept of the scope of an outsourcing arrangement and how it is structured and carried out (ISO 37500: 2014)

Performance based payment system Method of payment based on agreed output criteria (BS EN 15221-2:2006)

Performance Measurable result. Note 1: Performance can relate either to quantitative or qualitative findings. Note 2: Performance can relate to the management of activities, processes, products (including services), systems or organizations (ISO 41011:2017)

Performance requirement Performance criterion minimum acceptable level of a critical property (ISO 15686-1:2011)

Policy Intentions and direction of an organization as formally expressed by its top management (ISO 9000:2015)

Pre-design process Construction process determining construction properties for the built environment before it is designed (BS EN ISO 12006-2:2020)

Principle Fundamental basis for decision making or behaviour (ISO 26000:2010)

Procedure Specified way to carry out an activity or a process. Note 1: Procedures can be documented or not (ISO 9000:2015)

Process Set of interrelated or interacting activities that use inputs to deliver an intended result Note 1: Whether the "intended result" of a process is called output, product or service depends on the context of the reference. Note 2: Inputs to a process are generally the outputs of other processes and outputs of a process are generally the inputs to other processes. Note 3: Two or more interrelated and interacting processes in series can also be referred to as a process. Note 4: Processes in an organization are generally planned and carried out under controlled conditions to add value (ISO 9000:2015)

Procurement Activity of acquiring goods or services from suppliers. Note 1: The procurement process considers the whole cycle from identification of needs through to the end of a services contract or the end of the life of goods, including disposal. Note 2: Sourcing is a part of the procurement process that includes planning, defining specifications and selecting suppliers (ISO 20400:2017)

Product Result of a process. Note 1: There are four generic product categories, as follows: services; software; hardware; processed materials. Note 2: Products can be tangible or intangible (ISO/IEC 17065:2012)

Provider (or supplier) Organization that provides a product or a service. Example: Producer, distributor, retailer or vendor of a product or a service. Note 1: A provider can be internal or external to the organization. Note 2: In a contractual situation, a provider is sometimes called "contractor" (ISO 9000:2015)

Qualitative indicator Measure derived from a checklist or descriptive scale without any quantification. Note: qualitative indicators can be derived from checklists or descriptive scales or categorized into classes that can be assigned numeric values (ISO/DIS 59020:2023)

Quality Totality of features and characteristics of a product or service that bears on the ability of the product or service to satisfy stated or implied needs (ISO 10845-1:2020)

Quality policy Policy related to quality. Note 1: Generally the quality policy is consistent with the overall policy of the organization, can be aligned with the organization's vision and mission and provides a framework for the setting of quality objectives. Note 2: Quality management principles presented in this International Standard can form a basis for the establishment of a quality policy (ISO 9000:2015)

Quantitative indicator Measure based on numeric data that can be used for mathematical calculations and statistical analysis. Note 1: the input data can be directly measured or otherwise obtained. Note 2: quantitative input data are based on a physical or economic unit of measurement (ISO/DIS 59020:2023)

Real Estate Encompasses land along with anything affixed to the land, such as buildings. Note: Real estate, immovable property, real property, realty are used synonymously (BS EN 15221-4:2011). Immoveable property including structures, grounds and undeveloped land (ISO 41011:2017)

Recyclability Ability of component parts, materials or both to be separated and reprocessed from products and systems and subsequently used as material input for the same or different use or function (ISO 20887:2020)

Recyclable Characteristic of a product or associated component that can be diverted from the waste stream through available processes and programmes and can be collected, processed and returned to use in the form of raw materials or products. Note 1: whilst many products, components and materials are technically recyclable, in practice, recycling facilities might not be readily available or economically feasible to use. Note 2: recycling infrastructure for the material should exist in at least 60% of locations where the product is sold (ISO 14021:2016)

Recycling Activities to obtain recovered resources for use in a process or a product, excluding energy recovery. Note 1: Activities to obtain resources include activities such as recovery, collection, transport, sorting, cleaning, and re-processing. Note 2: Recycling does not include reuse (ISO/DIS 59004:2023)

Recover value Process of recuperating the value of the object of consideration (i.e., a resource) (ISO/DIS 59004:2023)

Recoverable resource Resource that can potentially be recovered and used again after it has already been processed or used (i.e. "post-consumption resource", "pre-consumer material", "post-consumer material"). Note 1: recovery can be undertaken to add or recover value. Note 2: a recoverable resource can provide no value and be considered waste (ISO/DIS 59004:2023)

Recovered resource or Secondary resource Resource that is obtained from a resource that has already been processed or used. Note 1: recovery can be undertaken to add or recover value. Note 2: a recovered resource may provide no value to the holder and be considered waste. Note 3: other terminology used, depending on the context, includes "secondary material" (ISO/DIS 59004:2023)

Refurbish Restore an item, during its expected service life, to a useful condition for the same purpose with at least similar quality and performance characteristics (ISO/DIS 59004:2023)

Refurbish-ability Ability to restore the aesthetic and functional characteristics of a product, building or other constructed asset to a condition suitable for continued use (ISO 20887:2020)

Refurbishing Process by which an item, during its expected service life, is restored to a useful condition for the same purpose and with at least similar quality and performance characteristics. Note 1: Refurbishing does not include restoration after the expected service life. Note 2: Refurbishing can include activities such as repair, rework, replacement of worn parts, and update of software or hardware but does not include activities that result in the need of a new certification of the product. Note 3: Refurbishing can result in a higher level of safety because safety updates released by the manufacturer for the relevant product are applied within refurbishment (ISO/DIS 59004:2023)

Registry system Framework of criteria useful for the classification and coding of spatial and technical elements of a building. It also includes an apparatus of standardized templates useful for collecting information, over time according to the same formalized scheme

Registry Database, structured according to the classification and coding rules defined by the Registry System, containing the information necessary to describe the consistency, functional and technical characteristics of a building

Remanufacturability Ability of a product to be disassembled and refabricated at the end of its useful life in a manner that provides restoration to a condition suitable for resale (ISO 20887:2020)

Remanufacture Return an item to original condition from both a quality and performance perspective using an industrial process (ISO/DIS 59004:2023)

Repair Returning a product, component, assembly, or system to an acceptable condition by renewal or replacement of worn, damaged, or degraded parts (ISO 20887:2020). Action to restore a product to a condition needed for the product to function according to its original purpose Note: Actions can include renewal or replacement of worn, damaged, or degraded parts of the product (ISO/DIS 59004:2023)

Replacement Change of parts of an existing item to regain its functionality (ISO 20887:2020)

Repurpose Adapt a product, or its component parts for use in a different function than it was originally intended for without making major modifications to its physical or chemical structure (ISO/DIS 59004:2023)

Repurposing Process by which a product, or its component parts are adapted for use in a different function than it was originally intended for without making major modifications to its physical or chemical structure (ISO/DIS 59004:2023)

Requirement Need or expectation that is stated, generally implied or obligatory Note 1: "Generally implied" means that it is custom or common practice for the organization and interested parties that the need or expectation under consideration is implied. Note 2: A specified requirement is one that is stated, for example in documented information. Note 3: A qualifier can be used to denote a specific type of requirement, e.g. product requirement, quality management requirement, customer requirement, quality requirement. Note 4: Requirements can be generated by different interested parties or by the organization itself. Note 5: It can be necessary for achieving high customer satisfaction to fulfill an expectation of a customer even if it is neither stated nor generally implied or obligatory (ISO 9000:2015)

Resource Asset from which a solution is created or implemented. Note 1: asset refers to physical resources such as natural resources, virgin resources, recoverable resources, and recovered resources. Note 2: a resource can be either a renewable resource or non-renewable resource. Note 3: resource includes any energy type, e.g. the energy content or energy potential of materials. Note 4: resources can be considered concerning both stocks and flows. Note 5: depending on the context, reference to 'resource' includes "raw material", "feedstock", "material" or "component." (ISO/DIS 59004:2023)

Retain value Process of maintaining the value of the object of consideration (i.e., a resource within the circular flow) (ISO/DIS 59004:2023)

Reusability Ability of a material, product, component or system to be used in its original form more than once and maintain its value and functional qualities during recovery to accommodate reapplication for the same or any purpose (ISO 20887:2020)

Reuse Use of products or components more than once for the same or other purposes without reprocessing. Note: reprocessing does not include preparation for re-use, such as removal of connectors, cleaning, trimming, stripping of coatings, packaging, etc. (ISO 20887:2020). Use of a product after its initial use, for the same purpose for which it was originally designed Note 1: Utilization intended by the original design can involve either single-use or multiple-uses by the initial user or customer over time. Note 2: Minor treatment (e.g. cleaning) of the product can be needed by the user to allow for reuse. Note 3: In some cases, resources, like water, are considered as a product. In these cases, design is not applicable (ISO/DIS 59004:2023)

Reverse logistics Process of managing, collecting, and moving products from their current location after the end-of-use for the purpose of recovering (3.1.7) or retaining value through proper handling. Note 1: End-of-use can occur when the organization (e.g. shop or retailer) in possession of the product considers it to be waste since it hasn't or can't be sold. Note 2: Only activities needed for the proper handling are included, e.g. logistics needed to bring a used product to a new customer are not included. Note 3: The proper handling can include remanufacturing, repair or recycling or other treatment (ISO/DIS 59004:2023)

Reversible connection Connection that can be disconnected and/or disassembled for easy alterations and additions to structures. Note 1 to entry: This is applicable to components, assemblies, modules or systems within a constructed asset (ISO 20887:2020)

Risk Effect of uncertainty on objectives. Note 1: An effect is a deviation from the expected—positive and/or negative. Note 2: Objectives can have different aspects (such as financial, health and safety, and environmental goals) and can apply at different levels (such as strategic, organization-wide, project, product and process). Note 3: Risk is often characterized by reference to potential events and consequences, or a combination of these. Note 4: Risk is often expressed in terms of a combination of the consequences of an event (including changes in circumstances) and the associated likelihood of occurrence. Note 5: Uncertainty is the state, even partial, of deficiency of information related to, understanding or knowledge of an event, its consequence, or likelihood (ISO Guide 73:2009) Risk is usually expressed in terms of risk sources, potential events, their consequences and their likelihood (ISO 31000:2018)

Risk assessment Overall process of risk identification, risk analysis and risk evaluation (ISO Guide 73:2009)

Risk management policy Statement of the overall intentions and direction of an organization related to risk management (ISO Guide 73:2009)

Risk management process Systematic application of management policies, procedures and practices to the activities of communicating, consulting, establishing the context, and identifying, analyzing, evaluating, treating, monitoring and reviewing risk (ISO Guide 73:2009)

Risk management Coordinated activities to direct and control an organization with regard to risk (ISO 31000:2018)

Risk owner Person or entity with the accountability and authority to manage a risk (ISO Guide 73:2009)

Risk source Element which alone or in combination has the potential to give rise to risk (ISO 31000:2018)

Sensor Device that observes and measures a physical property of a natural phenomenon or process and converts that measurement into a signal (ISO 23247-1:2021)

Service Level Agreement (SLA) Document which has been agreed between the demand organization and a service provider on performance, measurement and conditions of service delivery (ISO 41011:2017). Agreement between the client or customer and the service provider on performance, measurement and conditions of services delivery. A Facility Management agreement consists of general clauses, applicable to the whole agreement, and SLA specific clauses, only applicable to a facility service. In a Facility Management agreement, several SLAs are included (BS EN 15221-1:2006)

Service life Period of time after installation during which a facility or its component parts meet or exceed the performance requirements (ISO 15686-1:2011)

Service life planning Design process of preparing the brief and the design for the building and its parts to achieve the design life. Note 1: service life planning can, for example, reduce the costs of building ownership and facilitate maintenance and refurbishment (ISO 15686-1:2011)

Service provider Organisation that is responsible for the delivery of one or more facility services. Note: a service provider can be internal or external to the client's organisation (BS EN 15221-1:2006)

Service Output of an organization with at least one activity necessarily performed between the organization and the customer. Note 1: the dominant elements of a service are generally intangible. Note 2: service often involves activities at the interface with the customer to establish customer requirements as well as upon delivery of the service and can involve a continuing relationship such as banks, accountancies or public organizations, e.g. schools or hospitals. Note 3: Provision of a service can involve, for example, the following: (i) an activity performed on a customer-supplied tangible product (e.g. a car to be repaired); (ii) an activity performed on a customer-supplied intangible product (e.g. the income statement needed to prepare a tax return); (iii) the delivery of an intangible product (e.g. the delivery of information in the context of knowledge transmission); (iv) the creation of ambience for the customer (e.g. in hotels and restaurants); Note 4: A service is generally experienced by the customer (ISO 9000:2015)

Social responsibility Responsibility of an organization for the impacts of its decisions and activities on society and the environment, through transparent and ethical behaviour that: (i) contributes to sustainable development, including health and the welfare of society; (ii) takes into account the expectations of stakeholders; (iii) is in compliance with applicable law and consistent with international norms of behaviour; and (iv) is integrated throughout the organization and practised in its relationships. Note: Activities include products, services and processes (I SO 26000:2010)

Specification Detailed description of the essential performance and/or technical requirements for services or products and processes set out by the demand organization to make clear to the service provider the requirements to be fulfilled. Note: This is the documentary interface between the needs of the demand organization and the activities of the service provider (ISO 41011:2017). Document stating requirements (ISO 9000:2015)

Stakeholder Person or organization that can affect, be affected by, or perceive themselves to be affected by a decision or activity. Note 1: The term "interested party" can be used as an alternative to "stakeholder" (ISO 31000:2018)

Strategy Plan to achieve a long-term or overall objective (ISO 9000:2015)

Supplier Provider of a facility service or a product (BS EN 15221-1:2006)

Supply chain System of organizations, people, activities, information, and resources involved in delivering a product or service to an end user from a supplier (ISO 41011:2017). Sequence of activities or parties that provides goods or services to the organization (ISO 20400:2017). Series of processes or activities involved in the production and distribution of a material or product through which it passes from the source. Note: a supply chain is typically composed of a series of different organizations (ISO 22095:2020)

Sustainability State of the global system, including environmental, social and economic aspects, in which the needs of the present are met without compromising the ability of future generations to meet their own needs. Note 1: The environmental, social and economic aspects interact, are interdependent and are often referred to as the three dimensions of sustainability. Note 2: Sustainability is the goal of sustainable development (ISO Guide 82:2019)

Sustainable development Development that meets the environmental, social and economic needs of the present without compromising the ability of future generations to meet their own needs (ISO Guide 82:2019)

Sustainable procurement Procurement that has the most positive environmental, social and economic impacts possible over the entire life cycle. Note 1: Sustainable procurement involves the sustainability aspects related to the goods or services and to the suppliers along the supply chains. Note 2: Sustainable procurement contributes to the achievement of organizational sustainability objectives and goals and to sustainable development in general (ISO 20400:2017)

Taxonomy Practice and science of classification. Note: A knowledge map of a topic typically realised as a controlled vocabulary of terms and or phrases. An orderly classification of information according to presumed natural relationships. A classification system for improved information management, which should contribute to improving the capability of users to sustain and improve the operations of their business, into a series of hierarchical groups to make them easier to identify, study, or locate (BS EN 15221-4:2011)

Tender data Document that establishes the tenderer's obligations in submitting a tender and the employer's undertakings in administering the tender process and evaluating tender offers (ISO 10845-1:2020)

Tender offer Written offer to supply goods, carry out construction works and/or provide services under given conditions, usually at a stated price, and which is capable of acceptance and conversion into a binding contract (ISO 10845-1:2020)

Tenderer Person or organization that submits a tender offer (ISO 10845-1:2020)

Tendering (or bidding) Process of obtaining tenders with the intention of forming a contract with one or more of the tenderers (ISO 6707-2:2017)

Thing With regard to the Internet of things, this is an object of the physical world (physical things) or the information world (virtual things), which is capable of being identified and integrated into communication networks (ITU-T Y.2060)

Top management Person or group of people who directs and controls an organization at the highest level. Note 1: Top management has the power to delegate authority and provide resources within the organization. Note 2: If the scope of the management system covers only part of an organization, then top management refers to those who direct and control that part of the organization (ISO 14001:2015)

Traceability Ability to trace the history, application, location or source(s) of a material or product throughout the supply chain (ISO 22095:2020). Ability to trace the history, application and location of that which is under consideration Note 1: When considering a solution, traceability can relate to the following: origin of products, process and service history, and distribution and location of the product. Note 2: When considering a resource, traceability can relate to the following: origin of resource (e.g. whether it is a virgin or recovered resource), process history, and distribution and location of the resource (ISO 9000:2015; ISO/DIS 59004:2023)

Traceability system Manual or electronic system that provides the ability to access any or all information relating to the material or product under consideration throughout their life cycle, by means of accessing documented information. Note 1: "Life cycle" should be understood in the broadest possible sense, to include, for example, raw material extraction, agricultural production, final disposal, and reuse or recycling, as well as all other stages connected with product manufacture and use (ISO 22095:2020)

Unit of measurement Particular quantity, defined and adopted by convention, with which other quantities of the same kind are compared in order to express their magnitude relative to that quantity (ISO/IEC 15939:2017)

Useful life Time interval from first use until the instant when a limiting state is reached Note 1: The limiting state may be a function of failure rate, maintenance support requirement, physical condition, economics, age, obsolescence, changes in the user's requirements or other relevant factors. Note 2: The limiting state may be redefined by changes in conditions of use. Note 3 to entry: In this context, "first use" excludes testing activities prior to hand-over of the item to the end-user (BS EN 13306:2017)

Value Gain(s) or benefit(s) from satisfying needs and expectations, in relation to the use and the conservation of resources (e.g. revenues, savings, productivity, sustainability, satisfaction, empowerment, engagement, experience, public health, trust). Note 1: the gain can relate to the specific function and performance of a solution. Note 2: value is *relative to*, and *determined by* the perception of the interested party(ies). Note: value can be financial or non-financial e.g. social, environmental, and other gains or benefits. Note 4: value is dynamic over time (ISO 56000:2020; ISO/DIS 59004:2023)

Value chain Set of organizations that provide a solution that results in value for them (ISO/DIS 59010:2023)

Value network Network of interlinked value chains and interested parties (ISO/DIS 59010:2023)

Virgin resource Primary resource natural resource or energy that is used as a resource for the first time as input in a process or for creating a product. Note 1: virgin resources can be either a renewable resource or non-renewable resource. Note 2: other terminology used, depending on the context, includes "virgin material" or "primary material" (ISO/DIS 59004:2023)

Waste Resource that is considered to no longer be an asset as it, at the time, provides no value to the holder. Note 1: the holder can choose to retain, discard, or transfer the waste. Note 2: value can be assigned to waste as a result of a need from other interested parties, at which point the resource is no longer considered waste. Note 3: the assignment of value to waste as a resource is linked, in part, to the available technology (e.g. landfill mining). Note 4: some regulations require the holder to dispose of certain types of wastes, while others assign value to waste. Note 5: because resource includes the energy content or energy potential of materials, such energy, when liberated during a process and not recovered for another use, can be considered a waste (ISO/DIS 59004:2023)

Work result View of construction result by type of work activity and resources used (BS EN ISO 12006-2:2020)